SHAIVA PHILOSOPHY

SHAIVA PHILOSOPHY

Editor-in-Chief
Swami Chidatman Jee Maharaj

ANMOL PUBLICATIONS PVT. LTD.
NEW DELHI - 110 002 (INDIA)

ANMOL PUBLICATIONS PVT. LTD.
H.O.: 4374/4B, Ansari Road, Daryaganj,
New Delhi-110 002 (India)
Ph.: 23278000, 23261597
B.O.: No. 1015, Ist Main Road, BSK IIIrd Stage
IIIrd Phase, IIIrd Block,
Bangalore - 560 085 (India)
Visit us at: www.anmolpublications.com

Shaiva Philosophy

First Published, 2009
ISBN 978-81-261-3622-3

PRINTED IN INDIA

Printed at Mehra Offset Press, Delhi.

Contents

Preface

India colourful and vibrant, a land as diverse as its people. A mosaic of faiths, cultures, customs and languages that blend harmoniously to form a composite whole. One of the world's oldest living civilizations—which gave to the world—the concept of zero, the primordial sound Aum..., Yoga and Buddhism.

She is old. She is young. The two diametrically opposite, different yet unique and dynamic faces of India coexist even at this turn of the millennium. With a history of heritage, learning and civilisation dating back to almost 5000 years or perhaps even more, the modern India presents itself as an ideal country. From snow clad mountains to breezy sparkling blue sea side. From arid deserts with shifting sand dunes to lush green country sides. From ancient relics and architectural ruins dating back into antiquity to modern cities, India has it all.

To the ncrth it is bordered by the world's highest mountain chain, where foothill valleys cover the northernmost of the country's maximum states. Further south, plateaus, tropical rain forests and sandy deserts are bordered by palm fringed beaches. Its southernmost tip is the meeting point of three Oceans.

India is a land of staggering contrast with a mingling of the tradition and modernity that is an unique experience to be in India. One will forever remember the stay in this wonderful Homeland-a land truly of mystery and enchantment. Only India can offer such an astonishing variety of contrasts.

She has given birth to numerous spiritual leaders and founders of some of the popular religions of the present day world. India probably has the most religious diversity in any country. It's the birthplace of Hinduism, Buddhism, Sikhism and Jainism and is among the few places to have a resident

Zoroastrian population. India with her veritable treasure trove of culture, mysticism, philosophy, art, music and architecture to name a few. Much of India's classical music is devotional and the North Indian Hindustani and South Indian Carnatic streams are distinct and both have a complex 'raga' framework. The legacy of dance in India is tremendous. The classical dances of India are numerous such as Kathakali of Kerala, Bharatnatyam of Tamil Nadu, Kuchipudi of Andhra Pradesh, Manipuri and Odissi from Orissa are the prominent dance forms in this country that sways to an altogether novel beat. The Yakshagana, nautanki and puppetry are ancient folk forms that live on till date. The earliest specimens of Indian painting are the ones on the walls of the Ajanta Caves dating back to 2nd century BC. The Mughals had a huge impact on Indian art. The influence of Persian art brought placid garden scenes, illustrations from myths, legends and history into Indian art. Word craft, handicrafts, architecture and sculpture all contribute to this rich and varied domain. Indian literature, both in English and in the vernacular, is ever more popular around the world.

The book aims at presenting perspectives of different aspects of Indian culture. It has topics on Indian classical music and dance, painting, ayurveda and yoga which are traditional and indigenous system of physical and mental wellbeing, language, philosophy and excellence achievements by Indian in the different branches of science and social sciences. It also includes our ancient religious and classic literature.

It is hoped that this book will be welcomed by the general readers as also scholars who are interested in the Indian heritage.

—Editors

1

Shiva and Shaivism

History of Shaivism

The worship of Shiva is a pan-Hindu tradition, practiced widely across all of India. Shaivism has many different schools showing both regional variations and differences in philosophy. Shaivism has a vast literature that includes texts representing multiple philosophical schools, including non-dualist (abheda), dualist (bheda), and non-dual-with-dualism (bhedbheda) perspectives.

It is very difficult to determine the early history of Shaivism. Axel Michaels explains the composite nature of Shaivism as follows:

> *Like Vishnu, Shiva is also a high god, who gives his name to a collection of theistic trends and sects: Shaivism. Like Vaishnavism, the term also implies a unity which cannot be clearly found either in religious practice or in philosophical and esoteric doctrine. Furthermore, practice and doctrine must be kept separate.*

Rudra and Shiva

Shaivism, also known as the Shaiva cult, is devoted to worship of the god Shiva. The Sanskrit word is an adjective meaning kind, friendly, gracious, or auspicious. As a proper name it means "The Auspicious One", used as a euphimistic name for Rudra. In simple English transliteration it is written either as *Shiva* or *Siva*. Over the course of time, many regional cults were amalgamated into the figure of Shiva as we know him today.

Ancient Artifacts

Some people believe that artifacts from Mohenjadaro, Harappa and other archaeological sites of northwestern India and Pakistan indicate that some early form of Shiva worship was practiced in the Indus Valley. These artifacts include lingams and the "Pasupathy seal" that has been the subject of much study. The Indus Valley civilization reached its peak around 2300-2000 BC, when trade links with Mesopotamia are known to have existed, was in decline by 1800 BC, and faded away by 1500 BC.

Pasupathy Seal

A seal discovered during excavation of the Mohenjadaro archaeological site in the Indus Valley has drawn attention as a possible representation of a "proto-Shiva" figure. This "Pasupathy" (Lord of Animals, Sanskrit *pasupati*) seal shows a large central figure that is surrounded by animals. The central figure is often described as a seated figure, possibly ithyphallic, surrounded by animals. Sir John Marshall and others have claimed that this figure is a prototype of Shiva, and have described the figure as having three faces, seated in a "yoga posture" with the knees out and feet joined. Semi-circular shapes on the head are often interpreted as two horns. Gavin Flood characterizes these views as "speculative", saying that while it is not clear from the seal that the figure has three faces, is seated in a yoga posture, or even that the shape is intended to represent a human figure, it is nevertheless possible that there are echoes of Shaiva iconographic themes, such as half-moon shapes resembling the horns of a bull.

Worship

In the Temples: "Shivacharyas" conduct Shiva worship services. Only the Shivacharyas may enter the sanctum sanctorum, while worshippers gather at the entrance to watch. Unlike Catholic priests, Shivacharyas are dedicated solely to worship and do not perform marriages or other civil rites of passage. In Chidambaram and a few other places adhishaivas are allowed to perform the ceremonies. Services are held daily, as many as six each day depending on the resources and the popularity of the temple. The usual service consists of the following: first, the figure of the deity is anointed with oil, water, milk, ghee, honey, curd, various juices, sandalwood paste, and others before being showered with blossoms. Then it is dressed in the traditional way of Tamilnadu, adorned with jewels and flower garlands.

Incense is burned, followed by a food offering (usually a rice preparation). Beautiful lamps of various designs are lit and presented to the image of the deity. Camphor is lit and presented. The burning camphor is then carried to the congregation. The worshippers reverentially show their palms over it before placing their palms over their eyes, some say this gesture signifies that the devotion is as precious to the worshipper as his or her own sight. Finally sacred ash and kungumam are distributed into the upraised palms of the worshippers, who touch it onto their foreheads. The worshippers then process along the circumambulation at least once before bowing low in prayer before the sanctum sanctorum, singing and reciting verses from the Vedic hymns, the Thevaram and others.

At Home: People also worship Shiva at home. They have natural lingam-shaped stones to which they perform ablution flower-worship and Nivedhanam. (food offering)

Salagramam is a black glossy stone naturally occurring in the bed of the Gandaki river, a tributary of the Ganges in Nepal. Salagarmam stones vary in shape, from cylindrical with rounded ends to flattened globes, and depending on the shape, the stones represent Shiva, Vishnu and others. While very big stones are chiselled into figures of gods for worship in temples, small ones are used to create personal altars for worship in the home.

Beyond India: It left a major imprint on the intellectual life of classical Cambodia, Champa in what is today south Vietnam, Java, Kashmir and the Tamil land. The wave of Saivite devotionalism that swept through late classical and early medieval India redefined Karnataka and Tamil Nadu.

Shaivite worship legitimized several ruling dynasties in pre-modern India be they the Chola, the Rajput or tribal. A similar trend was witnessed in early medieval Indonesia with the Majapahit empire and pre-Islamic Malaya.

Developmental Milestones

Pre-Gupta Period: The documentation of formal religious history, as opposed to archaeological evidence or scriptural mentions, is marked by Gavin Flood's remark that:

> *The formation of Shaiva traditions as we understand them begins to occur during the period from 200 BC to 100 AD.*

The two great epics of India, the Mahabharata and the Ramayana, deal extensively with stories of both Shiva and Vishnu, and there are references to early Shiva ascetics in the Mahabharata. The *Zvetsvatara Upanishad* (400-200 BC) is the earliest textual exposition of a systematic philosophy of Shaivism. As explained by Gavin Flood, the text proposes: a theology which elevates Rudra to the status of supreme being, the Lord who is transcendent yet also has cosmological functions, as does Shiva in later traditions. In the grammarian Patanjali's "Great Commentary" (Sanskrit: *Mahbhasya*) on Panini's Sanskrit grammar (second century BC), he describes a devotee of Shiva as clad in animal skins and carrying an iron lance as the symbol of his god, perhaps a precursor of Shiva's trident.

Gupta Period: Puranic Shaivism: It is with the Puranas that Shaivism spread rapidly, eventually throughout the subcontinent, through the singers and composers of the Puranic narratives. The Puranic literature developed during the Gupta dynasty (c. 320-500 AD) along with Smarta Brahmin forms of worship. The convergence of various Shaiva and Vaishnava trends, as well as their growing popularity, may have been partly the outcome of dominant dynasties like the Guptas assimilating the resources and cultural elements of their conquered territories.

The bulk of the material contained in the Puranas was established during the reign of the Guptas, with incremental additions taking place to the texts up to later medieval times.

There are eighteen major Puranas, and these are traditionally classified into three groups of six each, with Shiva considered to be the central deity in the Shiva Purana, Linga Purana, Matsya Purana, Kurma purana, Skanda Purana, and Agni Purana. However this traditional grouping is inexact, for while the Shiva Purana is strongly sectarian in its focus on Shiva, others are not so clearly sectarian and include material about other deities as well, particularly Vishnu.

The Puranic corpus is a complex body of materials that advance the views of various competing cults, as Gavin Flood explains: Although these texts are related to each other, and material in one is found in another, they nevertheless each present a view of ordering of the world from a particular perspective. They must not be seen as random collections of old tales, but as highly selective and crafted expositions and presentations of worldviews and soteriologies, compiled by particular groups of Brahmins to propagate a particular vision, whether it be focused on Vishnu, Shiva, or Dev+, or, indeed, any number of deities.

For example, the Vishnu Purana (4th century CE) presents a Vaishnava viewpoint in which Vishnu awakens, becomes the creator god Brahma to create the universe, sustains it, and then destroys it as Rudra (Shiva).

6th-7th Century AD: Shaiva theism was expounded in the Agamas, which number two hundred including the Upagamas (the "Lesser" Agamas), which were composed before the 7th century AD. In the 7th century AD, Banabhatta included the worship of Shiva in his account of the prominent religious sects of that time. In the 7th century AD the great Chinese traveller Xuanzang (Huen Tsang) toured India and wrote in Chinese about the prevalence of Shiva worship at that time, describing Shiva temples at Kanoj, Karachi, Malwa, Gandhar (Kandahar), and especially at Varanasi (Benares) where he saw twenty large temples dedicated to Shiva.

Adi Shankara's Impact

Smartism is a denomination of Hinduism that places emphasis on a group of five deities rather than just a single deity. The "worship of the five forms" (pancyatana pkj) system, which was popularized by the philosopher Adi Shankara (also known as Zaekaracharya) (between 650 and 800 AD, traditionally 788–820 CE) among orthodox Brahmins of the Smarta tradition, invokes the five deities Shiva, Ganesha, Vishnu, Devi, and Surya.

This system was instituted by Zaekaracharya primarily to unite the principal deities of the five major sects on an equal status. The monistic philosophy preached by Zaekaracharya made it possible to choose one of these as a preferred principal deity and at the same time worship the other four deities as different forms of the same all-pervading Brahman.

Shankara's Dig-vijaya mentions six Shaivite sects that were in existence at his time, but their existence in an organized way is not clearly established.

Shaiva Siddhanta

Shaiva Siddhanta is the oldest, most vigorous and extensively practiced Shaivaite Hindu school active today, encompassing millions of devotees, thousands of active temples and dozens of living monastic/ascetic traditions. It is the formalized theology of the divine revelations contained in the twenty-eight Shaiva Agamas written in Sanskrit.

The Shaiva Siddhanta tradition provides a fundamental basis for the ritual and theology of all other Shaiva groups within its heritage. It is the most important normative form of Shaivism in South India. The tradition may have originated in Kashmir, where it developed a sophisticated theology propagated by theologians such as Sajyojoti, Bhatta, Narayanakantha, and his son Bhatta Ramakantha (c. 950-1000 AD). The Shaivasiddhantins date as early as the sixth century AD, with Kashmiri Shaivism clearly identifiable from the tenth century. By the eleventh century AD Shaiva Siddhanta had faded in Kashmir but developed in Tamil Nadu, where it exists to the present time. By the eleventh or twelfth century AD Shaiva Siddhanta was well-established in the South.

By the thirteenth century the Shaiva Siddhanta give theological and ritual expression to the devotional hymns of the

Nayanars. In Tamil Nadu the tradition incorporated an emotional devotion (*bhakti*) expressed in the hymns of these Tamil saints.

The Shaiva Siddhanta is a dualist system, maintaining that there is a distinction between the Lord (Sanskrit *+svara*) and the soul, in contrast to the monistic Kashmir Shaivism which viewed the Lord and the soul as one.

Shaiva Siddhanta is considered an orthoprax Shaiva tradition. In contrast to the orthoprax Shaiva Siddhanta, a separate major division is represented by the Bhairava Tantras and their various subdivisions.

Nayanars

By the seventh century CE, the Nayanars, a tradition of poet-saints in the bhakti tradition developed in South India with a focus on Shiva by the comparable to that of the Vaishnava Alvars.

Tirumular, also spelled (Tirumklr or Tirumklar) the author of the *Tirumantiram* (also spelled *Tirumandiram*) is considered by Tattwananda to be the earliest exponent of Shaivism in Tamil areas. Tirumular is dated as 7th or 8th century AD by Maurice Winternitz. The *Tirumantiram* is a primary source for the system of Shaiva Siddhanta, being the tenth book of its canon. The devotional poems of the Nayanars are divided into eleven collections together with a Tamil Purana called the *Periya Puranam*. The first seven collections are known as the *Thevaram* and are regarded by Tamils as equivalent to the Vedas. They were composed in the 7th century CE by Sambandar, Appar, and Sundarar.

The *Tiruvacakam* by Manikkavacagar is an important collection of hymns of which Sir Charles Eliot wrote, "In no literature with which I am acquainted, has the individual religious life, its struggles and dejections, its hopes and fears, its confidence and its triumph received a delineation more frank and more profound." The *Tiruvacakam* praises Shiva as belonging to the southern country yet worshipped by people of all countries.

Tamil Areas

There are numerous Shiva temples in Tamil Nadu, most located in the Thanjavur region which was a major part of the Chola empire between 800 and 1200 A.D. A particular branch of Shaivism, the philosophy of Siddhanta Shaivam, is particularly popular in southern India, Sri Lanka, Malaysia, Singapore and every other country where Tamils are living.

V.S. Pathak, in his book *Shaivism in Early Medieval India as known from Archaeological Sources: Rajendra Chola* writes that Rajendra saw the best of the Shaivas in northern India when he went to worship at the Ganges, and brought them to settle in his own country in Kanchi.

Gurus

The first known guru of the Suddha, or "pure," Shaiva Siddhanta tradition was Maharishi Nandinatha of Kashmir (ca 250 BC), recorded in Panini's *book of grammar* as the teacher of Rishis Patanjali, Vyaghrapada and Vasishtha. The only surviving written work of Maharishi Nandinatha is the twenty-six Sanskrit verses, called the *Nandikesvara Kasika*, in which he carried forward the ancient teachings. Due to his monistic approach, Nandinatha is often considered by scholars as an exponent of the Advaita school.

The next prominent guru on record is Rishi Tirumular, a Siddha in the line of Nandinatha who came from the Valley of Kashmir to South India to propound the sacred teachings of the twenty-eight Shaiva Agamas. In his work the *Tirumantiram*, "Sacred Incantation," Tirumular put the vast writings of the Agamas and the Suddha Siddhanta philosophy into the Tamil language for the first time. Tirumular's Suddha Shaiva Siddhanta shares common distant roots with Mahasiddhayogi Gorakshanatha's Siddha Siddhanta in that both are Natha teaching lineages. Tirumular's lineage is known as the Nandinatha Sampradaya, while Gorakshanatha's is called the Adinatha Sampradaya.

Saints and Ascetics

Shaiva Siddhanta flowered in South India as a forceful bhakti movement infused with insights on siddha yoga. During the seventh to ninth centuries, saints Sambandar, Appar and Sundarar pilgrimaged from temple to temple, singing soulfully of Shiva's greatness. They were instrumental in successfully defending Shaivism against the threats of Buddhism and Jainism. Soon thereafter, a king's Prime Minister, Manikkavasagar, renounced a world of wealth and fame to seek and serve God. His heart-melting verses, called *Tiruvasagam*, are full of visionary experience, divine love and urgent striving for Truth. The songs of these four saints are part of the compendium known as *Tirumurai*, which along with the Vedas and Shaiva Agamas form the scriptural basis of Shaiva Siddhanta in Tamil Nadu.

Siddhas

Besides the saints, philosophers and ascetics, there were innumerable siddhas, "accomplished ones," God-intoxicated men who roamed their way through the centuries as saints, gurus, inspired devotees or even despised outcasts. Shaiva Siddhanta makes a special claim on them, but their presence and revelation cut across all schools, philosophies and lineages to keep the true spirit of Shiva present on Earth. These siddhas provided the central source of power to spur the religion from age to age. The well-known names include Agastya, Bhogar, Tirumular and Gorakshanatha. They are revered by the Siddha Siddhantins, Kashmir Shaivites and even by the Nepalese branches of Buddhism.

Shaiva Siddhanta Teachings

Rishi Tirumular, like his satguru, Maharishi Nandinatha, propounded a monistic theism in which Shiva is both material and efficient cause, immanent and transcendent. Shiva creates souls and world through emanation from Himself, ultimately reabsorbing them in His oceanic Being, as water flows into water, fire into fire, ether into ether.

The Tirumandiram unfolds the way of Siddhanta as a progressive, fourfold path of charya, virtuous and moral living; kriya, temple worship; and yoga—internalized worship and union with Para Shiva through the grace of the living satguru—which leads to the state of jnana and liberation. After liberation, the soul body continues to evolve until it fully merges with God—jiva becomes Shiva.

Affirming the monistic view of Shaiva Siddhanta was Srikumara (ca 1056), stating in his commentary, Tatparyadipika, on Bhoja Paramara's works, that Pati, pasu and pasa are ultimately one, and that revelation declares that Shiva is one. He is the essence of everything. Srikumara maintained that Shiva is both the efficient and the material cause of the universe.

The Spread of Shaiva Siddhanta

In Central India, Shaiva Siddhanta of the Sanskrit tradition was first institutionalized by Guhavasi Siddha (ca 675). The third successor in his line, Rudrasambhu, also known as Amardaka Tirthanatha, founded the Amardaka monastic order (ca 775) in Andhra Pradesh. From this time, three monastic orders arose that were instrumental in Shaiva Siddhanta's diffusion throughout India. Along with the Amardaka order (which identified with one of Shaivism's holiest cities, Ujjain) were the Mattamayura Order, in the capital of the Chalukya dynasty, near the Punjab, and the Madhumateya order of Central India. Each of these developed numerous suborders, as the Siddhanta monastics, full of missionary spirit, used the influence of their royal patrons to propagate the teachings in neighbouring kingdoms, particularly in South India. From Mattamayura, they established monasteries in the regions now in Maharashtra, Karnataka, Andhra and Kerala (ca 800).

Of the many gurus and acharyas that followed, spreading Siddhanta through the whole of India, two siddhas, Sadyojyoti and Brihaspati of Central India (ca 850), are credited with the systematization of the theology in Sanskrit. Sadyojyoti, initiated by the Kashmir guru Ugrajyoti, propounded the Siddhanta philosophical views as found in the Raurava Agama. He was succeeded by Ramakantha I, Srikantha, Narayanakantha and Ramakantha II, each of whom wrote numerous treatises on Shaiva Siddhanta.

Later, King Bhoja Paramara of Gujarat (ca 1018) condensed the massive body of Siddhanta scriptural texts that preceded him into a one concise metaphysical treatise called

Tattvaprakasa, considered a foremost Sanskrit scripture on Shaiva Siddhanta. Shaiva Siddhanta was readily accepted wherever it spread in India and continued to blossom until the Islamic invasions, which virtually annihilated all traces of Siddhanta from North and Central India, limiting its open practice to the southern areas of the subcontinent.

A New Siddhanta

It was in the twelfth century that Aghora Shiva took up the task of amalgamating the Sanskrit Siddhanta tradition of the North with the Southern, Tamil Siddhanta. As the head of a branch monastery of the Amardaka Order in Chidambaram, Aghora Shiva gave a unique slant to Shaiva Siddhanta theology, paving the way for a new pluralistic school.

In strongly refuting any monist interpretations of Siddhanta, Aghora Shiva brought a dramatic change in the understanding of the Godhead by classifying the first five principles, or tattvas (Nada, Bindu, Sadashiva, Èsvara and Suddhavidya), into the category of pasa (bonds), stating they were effects of a cause and inherently unconscious substances. This was clearly a departure from the traditional teaching in which these five were part of the divine nature of God. Aghora Shiva thus inaugurated a new Siddhanta, divergent from the original monistic Shaiva Siddhanta of the Himalayas.

Despite Aghora Shiva's pluralistic viewpoint of Siddhanta, he was successful in preserving the invaluable Sanskritic rituals of the ancient Agamic tradition through his writings. To this day, Aghora Shiva's Siddhanta philosophy is followed by almost all of the hereditary Shivacharya (shaiva-brahmins) temple priests, and his paddhati texts on the Agamas have become the standard puja manuals. His Kriyakramadyotika is a vast work covering nearly all aspects of Shaiva Siddhanta ritual, including diksha, sauskaras, atmartha puja and installation of Deities.

A Dualistic Development

In the thirteenth century, another important development occurred in Shaiva Siddhanta when Meykandar wrote the twelve-verse Shivajnanabodham. This and subsequent works by other writers laid the foundation of the Meykandar

Sampradaya, which propounds a pluralistic realism wherein God, souls and world are coexistent and without beginning. Shiva is efficient but not material cause. They view the soul's merging in Shiva as salt in water, an eternal oneness that is also twoness. This school's literature has so dominated scholarship that Shaiva Siddhanta is often erroneously identified as exclusively pluralistic. In truth, there are two interpretations, one monistic and another dualistic, of which the former is the original philosophical premise found in pre-Meykandar scriptures, including the Upanishads.

Shaiva Siddhanta Today

Today Shaiva Siddhanta has sixty million Tamil Shaivites in South India and Sri Lanka (See Hinduism in Sri Lanka). Prominent Siddhanta societies, temples and monasteries also exist in a number of other countries. The United States island of Kauai, a part of Hawaii, is home to the Shaiva Siddhanta Church, an organization that promotes the union of worldwide Hindus, Shaivites and others, through a publication called Hinduism Today.

Kashmir Shaivism

Included within the broad umbrella of the various Hindu philosophies, Kashmir Shaivism is a school of Shaivism with a highly monistic vision; its essence is "to realize that God and individual are one". This school of Shaivism arose during the ninth century CE.

Legend of Kashmir Shaivism

A book by Swami Lakshmanjoo says that it is traditionally thought that Kashmir Shaivism began with the Sat Yuga 1.7 million years ago. As the philosophy of Kashmir Shaivism is deeply rooted in the Tantras, the lineage of Kashmir Shaivism begins with Shiva himself.

According to tradition, as the knowledge of the Tantras were lost by the time of Kali Yuga, Shiva took the form of Shrikanthanath at Mt. Kailash, where he fully initiated Durvasa Zci, into all forms of the Tantrika knowledge, including *abheda* (without differentiation), *bhedabheda* (with and without differentiation), and *bheda* (differentiated), as described in the *Bhairava Tantras*, *Rudra Tantras*, and *Shiva Tantras*, respectively. Durvasa Zci intensely meditated in the hope of finding an adequate pupil to initiate, but failed to do so. Instead, he created three "mind-born" sons, and initiated the first son, Tryambaka fully into the monistic *abheda* philosophy of the *Bhairava Tantras*; this is known as Kashmir Shaivism.

The Shiva Sutras

The first great initiate recorded in history of this spiritual path was Vashugupta (c. 875-925). Vashugupta formulated for the first time in writing the principles and main doctrines of this system. After Vashugupta, the esoteric literature of Kashmir Shaivism explosively blossomed and lasted for almost four centuries, during which it became very voluminous and it reached an exceptional philosophical depth.

A fundamental work of Shaivism, traditionally attributed to Vashugupta, is the Shiva Sutras of Vashugupta. Traditionally these sutras are considered to have been revealed to Vashugupta by Shiva.

According to myth, Vashugupta had a dream in which Shiva told him to go to the Mahdeva mountain in Kashmir. On this mountain he is said to have found verses inscribed on a rock, the *Shiva Sutras*, which outline the teachings of Shaiva monism. This text is one of the key sources for Kashmir Shaivism. The work is a collection of aphorisms. The sutras expound a purely non-dual (*advaita*) metaphysics. These sutras, which are classified as a type of Hindu scripture known as *agamas*, are also known as the *Shiva Upanishad Samgraha* (Sanskrit: *Sivopanicad saegraha*) or *Shivarahasyagama Samgraha*.

Classification of the Written Tradition

Some of the information in this article or section may not be verified by reliable sources. It should be checked for inaccuracies and modified to cite reliable sources.

The first Kashmiri Shaiva texts were written in the early ninth century CE.

As a monistic tantric system, Trika Shaivism, as it is also known, draws teachings from shrutis, such as the monistic *Bhairava Tantras*, Shiva Sutras of Vashugupta, and also a unique version of the *Bhagavad Gita* which has a commentary by Abhinavagupta, known as the *Gitartha Samgraha*. Teachings are also drawn from the *Tantraloka* of Abhinavagupta, prominent among a vast body of *smritis* employed by Kashmir Shaivism.

In general, the whole written tradition of Shaivism can be divided in three fundamental parts: *Agama Shastra*, *Spanda Shastra* and *Pratyabhijna Shastra*.

1. *Agama Shastra* are those writings that are considered as being a direct revelation from Shiva. These writings were first communicated orally, from the master to the worthy disciple. They include essential works such as *Malenivijaya Tantra*, *Svacchanda Tantra*, *Vijnana Bhairava Tantra*, *Ratnamala Tantra*, *Mrgendra Tantra*, *Rudrayamala Tantra*, *Shiva Sutra* and others. There are also numerous commentaries to these works, *Shiva Sutra* having most of them.
2. The main work of this category is *Spanda Karika* of Vashugupta, which has many commentaries. Out of these many commentaries, two are of major importance: *Spanda Sandoha* (this commentary talks only about the first verses of *Spanda Karika*), and *Spanda Nirnaya* (which is a commentary of the complete text).
3. *Pratyabhijna Shastra* are those writings which have mainly a metaphysical content. Due to their extremely high spiritual and intellectual level, this part of the written tradition of Shaivism is the least accessible for the uninitiated. Nevertheless, this corpus of writings refer to the simplest and most direct modality of spiritual realization. *Pratyabhijña* means "recognition" and refers to the spontaneous recognition of the divine nature hidden in each human being (atman). The most important works in this category are: *Isvara Pratyabhijna*, the fundamental work of Utpaladeva (who is a central figure of the Shaivism tradition – he

combined in a harmonious way the impetuous and endless love for the Divine and the most refined rational knowledge and metaphysical discernment), and *Pratyabhijna Vimarasini*, a commentary to *Isvara Pratyabhijna*. *Isvara Pratyabhijna* means in fact the direct recognition of the Heart (*i.e.*, essence), meaning the direct recognition of God-the-Father inside ourselves. Before Utpaladeva, his master Somananda wrote *Shiva Drsti* (*The Vision of Shiva*), a devotional poem written on multiple levels of meaning.

Lingayatism

History and Origin

The history of Lingayats spans around yugas. But as per current available historical evidences, we can see that Veerashaivism is existed long back, around 6th Century AD (There is ruling by Allahabad High court in 1920, Regarding Kashi Vishwaradhya peetha at Jangamwadimath Kashi).

The Basavanna (1134-1196 AD), It is said that Basavanna, though born a *Brahmin* rebelled against the rigid practices of the caste system then prevalent and eventually began expounding his own theosophy with a casteless society at its core. Soon, his philosophy began attracting large numbers of people into the fold. Saints like Allama Prabhu, Akka Mahadevi, Channabasavanna also played pivotal roles in spearheading the sect.

Basavanna was finance minister in Bijjal kindom. He lived and taught in the northern districts of what is now Karnataka State. He lived in Basavakallyan of bidar district major part of his life and Basavakallyan was known as his karmabhoomi. This movement found its roots during the brief rule of the southern Kalachuri dynasty in those parts of the state. Some believers believe that Basavanna is an incarnation of Nandi, Shiva's greatest devotee.

However, some scholars maintain that Basavanna only brought about a renaissance in an already existing sect. They attribute the founding of the sect to the *Panchacharyas*. The

Panchacharyas or the 'Five Prophets' consisted of Sri Revanaradhya or Revanasiddha, Marularadhya or Marulasidhdha, Ekoramaradhaya or Ekorama, Panditaradhya and Vishwaradhya.

Sources of Lingayat History

To reconstruct the historical context, origin and development of the Lingayat faith, one will need to draw upon several sources. The various sources that are involved here include folk literature, inscriptions, historical facts, post-Basava developments, the Vachanas, reminiscences, Ragale literature, the lingayat and Jaina puranas and the doctrinal and philosophical works like the *Shoonya Sampadane*. Following a holistic and comparative study of all these sources, it is possible to reconstruct the origin and development of the Lingayat faith across time and space. Of all the sources, particularly important is the epigraphical evidence. The Arjunavada inscription in particular establishes the historicity of Basava beyond dispute. The *Vachana* literature, especially the *vachanas* composed by Basava himself are of vital importance in this studya. Harihara's *Ragale* is yet another invaluable source.

Allahabad High court ruling in 1920, regarding Kashi peetha. Shasanagalli Panchacharyaru: A thesis for which Ph.D awarded to Dr. Rajashekhar Swamy Gorata by Bangalore University.

Lingayat Theosophy

Lingayats believe in a monotheistic world where Shiva the supreme God and self are one and the same. This form of monotheism is called *Shakti Vishishtadvaita*. However, unlike other Hindus, Lingayats reject the Vedas but rather focus more on the Hindu Agamas, specifically, the Shaivite Agamas. Some Lingayats view the Vedas to be polytheistic in nature while the Agamas are held as being strictly monotheistic and devotional in nature.

Veera Shaivism's means of attainment depends on the concepts of Panchachara (five codes of conduct) and ashtavarana (eight shields) and *Shatsthala* which are central to Lingayat theosophy.

The *Panchacharas* describe the five modes of conduct to be followed by the believer. The *Ashtavaranas* shield the devotee from extraneous distraction and worldly attachments. The *Shatsthala* refers to the different levels of attainment that the devotee can achieve, to protect the body as the abode of the Lord.

Panchacharas

Panchachara or the five modes of conduct of:

- Lingachara-daily worship of the personal Shivalinga
- Sadachara-attention to vocation and duty
- Shivachara-acknowledging Shiva as the one God and equality among members
- Bhrityachara-humility towards all creatures
- Ganachara defense of the community and its tenets

Ashtavarana

Ashtavarana or the 'eight protections' include: Guru, Linga, Jangama (wandering monk), Padodaka (water from bathing the Linga or guru's feet), Prasada (sacred offering), Vibhuti (holy ash created using cow dung), Rudraksha (holy beads) and Mantra (Namah Shivaya).

Shatsthala

'Shatsthala *or the concept of six phases/states/paths is pivotal to the Lingayat philosophy.* Shatsthala *is a conflation of* Shat *and* Sthala *which means 'six phases/states/levels' through which a soul advances in its ultimate quest of realisation of the Supreme. The Shatsthala comprises the* Bhakta Sthala, Maheshwara Sthala, Prasadi Sthala, Pranalingi Sthala, Sharana Sthala *and the* Aikya Sthala. *The* Aikya Sthala *is the culmination where the soul leaves the physical body and merges with the Supreme.* While the origins of the *Shatsthala* may be traced to the *Agamas* particularly the *Parameshwaratantra*, with the evolution of Veerashaivism, the evolution of concept of Shatsthala was also unavoidable. While Basavanna understood

Shatsthala as a process with various stages to be attained in succession, the credit of refuting this and redefining *Shatsthala* goes to Channabasavanna, Basavanna's nephew. Channabasavanna differed radically from his uncle and held that a soul can reach its salvation in any stage.

Concept of God

Lingayats believe in a monotheistic world where Shiva is the supreme God and the self and Shiva are one and the same. This form of monotheism is called *Shakti Vishishtadvaita*. The concept of God in Veerashaivism is very interesting and its synthesis is even more fascinating. The Istalinga worn by the Veerashaivas on their body is technically a miniature of the *Sthavaralinga* worshipped in the temples. The Sthavaralinga is also known as *Shivalinga* and it is considered to be an amorphous representation of Shiva.

Concept of Shoonya

True union and identity of Shiva (Linga) and soul (anga) is life's goal, described as shoonya, or nothingness, which is not an empty void. One merges with Shiva by shatsthala, a progressive six-stage path of devotion and surrender: bhakti (devotion), mahesha (selfless service), prasada (earnestly seeking Shiva's grace), pranalinga (experience of all as Shiva), sharana (egoless refuge in Shiva), and aikya (oneness with Shiva). Each phase brings the seeker closer, until soul and God are fused in a final state of perpetual Shiva consciousness, as rivers merging in the ocean.

Anubhava Mantapa

The *Anubhava Mantapa* was an academy of mystics, saints and philosophers of the Veerashaiva faith in the 12th century Kalyana. It was the fountainhead of all religious and philosophical thought pertaining to the Veerashaivas. It was presided over by the mystic Allama Prabhu and numerous Sharanas from all over Karnataka and other parts of India were participants. This institution was also the fountainhead of the Vachana literature which was used as the vector to propagate Veerashaiva religious and philosophical thought.

Other giants of Veerashaiva theosophy like Akka Mahadevi, Channabasavanna and Basavanna himself were participants in the Anubhava Mantapa. The Anubhava Mantapa was originally called the *Shivanubhavamantapa*. It is claimed that Anubhava Mantapa was first to introduced democracy to the world as early as 12 century.

Saguna and Nirguna

Saguna is one who has Guna or characteristic manifestations. This way leads to Saguna Upasane. This is accomplished by engaging our senses like eyes to watch Abhisheka, ears to hear Sthotra and Mantra, and tongue to chant etc. This means that involving our characteristics for worship.

Nirguna means one who has gone past all attributes. When a person follows the Ashtavarana for the body and Shatsthala for the mind (these gradually converts a person from Saguna to Nirguna) will lead self to rise and reach nirguna state. This way leads to Nirguna Upasane.

Scriptures

- Siddhanta Shikhamani
- Vachanas
- Mantra Gopya
- Shoonya Sampadane
- Shaivite Agamas
- Karana Hasuge

Lingayat Customs and Practices

Ishtalinga

The Lingayats make it a point to wear the *Ishtalinga* at all times. The Istalinga is made up of light gray slate stone coated with fine durable thick black paste of cow dung ashes mixed with some suitable oil to withstand wear and tear. Sometime it is made up of ashes mixed with clarified butter. The coating is called *Kanti* (covering). Though the *Ishtalinga* is sometimes likened to be a miniature or an image of the

Sthavaralinga, it is not so. The *Ishtalinga* on the contrary is considered to be Lord Shiva himself and its worship is described as *Ahangrahopasana*.

Thus, for the Lingayats it is an amorphous representation of God. Lingayat thus means the wearer of this Linga as *Ishta Linga*. Here the word *Ishta* is a Sanskrit term meaning 'adored' or 'desired'. Unlike Advaitins however, Lingayats do not treat the Ishtalinga as merely a representation of God to aid in realising God but worship the Ishtalinga itself as God. Lingayats do not partake non-vegetarian food like most brahmins.

Lingadharane

Lingadharane is the ceremony of initiation among Lingayats. Though, it can be performed at any age, it is usually performed when a child is 3-8 days old.

The child receives Istalinga from the family Guru that is worshipped until age 8-11 years. Usually between the ages of 8 and 11 years, the child receives *Diksha* from the Guru. From then on, the child wears the Linga at all times for the remainder of his/her life and it is worshipped as their own *Istalinga*.

The *Linga* is housed in a small silver and wooden box and cloth. It is worn on the chest or around the body using a thread. Unlike other castes of Hinduism like brahmins permitting upanayana to only males this ceremony or deeksha can be taken by both men and women in presence of a satguru.

This practice was started by revolutionary Basavanna himself who refused to undergo upanayana seeing the discrimination of women.

Kaayakave Kailash

Kaayaka means the exertion of the Kaaya (body) for the liberation of the soul imprisoned therein. Kailash means abode of Shiva-heavenly.

- *Kaayakave Kailash* literally means, Kaayaka or the body which exerts itself for nishkaama Karma-Karma without any expectation is nothing but Kailash-the abode of Shiva-heavenly.

A vachana complimentary to this which talks about keeping the Kaaya or body purely goes as

- *yenna kaale kamba dehave degula shirave honna kalashavayya sthaavakkalivuntu jangamakkalivilla*

As one theory goes Indian subcontinent can be divided as North and South divided by the Vindhya mountain ranges. While the North is blessed with the Himalayan rivers flowing all through the year and boasts of Ganga-called as sacred river. The South has river Kaveri which originates at Talakaveri dries up in summer. Hence the North is referred as Punya Bhoomi residents of which believe in taking a dip in Ganga with Bhakti will wash off all your sins. But South is referred as Karma Bhoomi who believe in their Karma which will decide the fate. The Kaayaka Tatva of Basavanna also bases itself into Karma Siddhantha-Philosophy of Karma.

Daasoha

Among the many injunctions prescribed for the devout Lingayat, *Daasoha* is a very important one. Basava created this as a protest against the feudalistic ideologies present at that time. He shunned the sharp hierarchial divisions that existed and sought to remove all distinctions between the hierarchically superior master class and the subordinate, servile class. Even though he himself served as a minister under the king, Bijjala, he pointed out that he worked only as a *daasohi* or one who serves.

Daasoha to him meant working hard for one's livelihood and for the maintenance of society. In Basava's view, a *daasohi* should consider himself, but a servant of society. Therefore, *Daasoha* in principle assumed that what belongs to God must return to Him and what came from society should be given back by way of selfless service. Basava exhorted all wearers of *Ishta linga* to practice *Daasoha* without reservation.

A famous vachana says:

- *Soham yennade Daasoham yendenisayya*-which means be selfless (Daasa Aham) rather than selfish (Naanu or Aham).

Burial

Unlike most other Hindus who cremate the dead, the Lingayat bury their dead. The dead are buried in the *Dhyana mudra* (meditating position) with their *Ishta linga* in their left hand.

Veerashaiva Literature

The rise of Veerashaivism heralded a new and glorious chapter in the annals of Kannada literature. It saw the birth of the Vachana style of literature with the Veerashaiva philosophy at its core.

The Vachanas were pithy poems of devotional nature that expounded the ideals of Veerashaivism. Saints and Sharanas like Allama Prabhu, Akka Mahadevi and Basavanna himself were at the forefront of this development. The entire corpus of these works was in Kannada and it marks one of the glorious chapters in the history of Kannada literature. As with the Dasa Sahitya of the later Haridasas, the Vachanas were also primarily targeted at the common man and sought to demystify God as large sections of society had been deprived of access to the texts. The Jangamas played a central role in the propagation of the Vachanas.

Veerashaiva Mathas and Mathadipathis

Historically there are 5 mathas known as pancha peethas located at:

- Balehonnuru: Rambhapuri Peetha
- Ujjaini-Kudligi, Bellary district, Karnataka state.
- Kedar-Uttaranchal
- Shreeshaila-Andhra Pradesh.
- Kashi-Uttar Pradesh.

Lingayat Demographics

Lingayats today are spread all over the state of Karnataka. Sizeable populations are also found in parts of Maharashtra and Andhra Pradesh bordering Karnataka. In Karnataka, the Lingayat population is quite large and widespread. Lingayats are spread in several villages such as Sogathurai, Katteri etc in the Nilgiri District of Tamil Nadu. They constitute roughly 15-20% of the total population of Karnataka.

Other than this, Lingayat diaspora can be found in countries around the world, particularly the United States, Britain and Australia. The Lingayat community wields considerable clout in the sphere of Karnataka politics. Several important and powerful leaders have been from the Lingayat community.

Lingayats and Social Work

The Lingayat community under the aegis of several Mathas has been very active in the field of social work, particularly in the field of education and medicine. Thousands of schools are run by the Lingayat *Mathas* where education, sometimes free and with boarding facilities is provided to students of all sections of society irrespective of religion or caste. In addition, various Lingayat organisations run numerous schools, colleges and hospitals across the length and breadth of Karnataka. Some of these institutions also have branches in other states of India. Some of the notable Lingayat run institutions include the JSS group of institutions and the K.L.E.

Famous Lingayats

- Dr. S. J. Nagalotimath (Internationally renowned doctor)
- Dr. M. C. Modi (Guinness Record holder for eye operation)
- Shivraj Patil (Present Home Minister, Government of India)
- M. P. Prakash (Present Home, Law, Parliamentary affairs Minister of Karnataka)
- B D Jatti (Former Vice President of India and former CM of Karnataka)
- S. Nijalingappa (Former CM of Karnataka)
- Basavaraj Rajguru (One of the best Classical Music Masters of Karnataka)
- Mallikarjun Mansur (Great Classical Musician)
- G. S. Shivarudrappa, poet and critic, Rashtra Kavi.

- B S Yediyurappa (Present Dy. CM of Karnataka)
- Baba Kalyani (Promoter of Bharat Forge)
- Late Veerendra Patil (former Chief minister, Karnataka and Union Minister, Govt. of India)
- Dr. M ChidanandaMurthy (Famous historian)
- Late Dr. D M Nanjundappa (Noted economist, Chairman of Committee for Redressal of Regional imbalances, Former Chairman Karnataka Planning Commission)
- S R Bommai (senior leader Janatha Dal and former CM Karnataka)
- Late J H Patel J. H. Patel (former CM Karnataka)
- Shankar Bidari (Senior IPS Officer, Karnataka Police)
- Prabhakar Kore (Chairman, KLE Society, Biggest education society in Asia)

Bhakti Movement

Bhakti movements are Hindu religious movements in which the main spiritual practice is the *fostering of loving devotion to God*, called bhakti. They are monotheistic movements generally devoted to worship of Shiva or Vishnu or Shakti.

Origins

The Bhakti Movement was essentially founded in South India and later spread to the North during the late medieval period. The notion of 'Bhakti' (loosely translated as devotional love to God) is of antiquity. A nascent consciousness of what 'Bhakti' constitutes is already to be found in the earliest Vedas, especially in relation to deities such as Varuna. A clearer expression of Bhakti began to be formed during the so-called Epic Period and the Puranic periods of Hindu history. Texts such as the Bhagavad Gita and the Bhagavata Purana clearly explore Bhakti Yoga or the Path of Devotion as a means to salvation.

The Bhakti Movement itself is a historical-spiritual phenomenon that crystallized in South India during Late Antiquity. It was spearheaded by devotional mystics (later revered as Hindu saints) who extolled devotion and love to God as the chief means of spiritual perfection. The Bhakti movement in South India was spearheaded by the sixty-three Nayanars (Shaivite devotees) and the twelve Alvars (Vaishnavaite devotees).

Among the earliest Shaivite mystics was Karaikkal Amaiyar, who probably lived around the late 5th century AD or perhaps the early 6th century. She was said to be a contemporary of the Vaishnavaite saints Bhuttalwar and Peialwar. Kannapa Nayanar was also an early Shaiva Bhakti saint. But most famous among the Shaiva Bhakti saints were the 'Nalvar' (The Four Eminent Ones), namely Sundarar, Appar, Sambandar and Manikkavasagar. Their devotional hymns are ecstatic, lyrical and moving.

The Vaishnavaite Bhakti movement was contemporaneous with the Shaiva Bhakti movement. The hymns of the twelve alvars are held together as the 'Nalayira Divya Prabandham' and recited (as are the Shaiva texts) in temple rituals. Whilst all the saints are held in great reverence, Aandaal (or Goda-devi) in particular holds a special place among the Vaishnava saints. Not only is she the only female Vaishnava saint but also her hymns are among the best expressions of bridal mysticism in the Hindu religion.

The twelve Alvars and the sixty-three Nayanars nurtured the incipient bhakti movement in South India under the Pallavas and Pandyas in the fifth to seventh centuries AD. They constitute South India's 75 Apostles of Bhakti and were greatly influential in determining the expression of faith in South India. The path of devotion as expounded by these mystics would later be incorporated into Ramanuja and Madhva philosophical systems.

During the 12th and 13th centuries A.D., the Virashaiva movement and, during the rule of the Vijayanagar Empire in South India, the Haridasa movement spread from present-day Karnataka. The Virashaiva movement spread the philosophy of Basavanna, a Hindu reformer. The seeds of Carnatic music were sown, and the philosophy of Madhvacharya was propagated by the Kannada Haridasas. The Haridasa movement presented, like the Virashaiva movement, another strong current of Bhakti, pervading the lives of millions.

The Haridasas presented two groups–Vyasakuta and Dasakuta. The former were required to be proficient in the Vedas, Upanishads and other Darshanas, while the Dasakuta merely conveyed the message of Madhvacharya through the Kannada language to the people. The philosophy of Madhvacharya was preserved and perpetuated by his eminent disciples like Vyasatirtha or Vyasaraja Naraharitirtha, Vadirajatirtha, Sripadaraya, Jayathirtha and others. In the fifteenth century, the Haridasa movement took shape under Sripadaraya of Mulbagal; but his disciple Vyasatirtha provided it a strong organizational base. He was intimately associated with the Vijayanagar Empire, where he became a great moral and spiritual force. His eminent disciples were Purandaradasa and Kanakadasa. Yakshagana, as a theatre form emerged as an offshoot of this movement in Karnataka.

The late Bhakti movement led to the proliferation of regional poetic literature in the various vernacular languages of India. The Bhakti movement in what is now Karnataka resulted in a burst of poetic Kannada literature in praise of Lord Vishnu. Some of its leaders include Purandara Dasa and Kanaka Dasa, whose contributions were essential to Carnatic music. The later Carnatic Trinity is also no doubt a product of this long Bhakti Movement.

The Bhakti movement began to spread to the North during the late medieval ages when North India was under Muslim domination. There was no grouping of the mystics into Shaiva and Vaishnava devotees as it was in the South. The movement was spontaneous and the various mystics had their own version of devotional expression. Unlike in the South where devotion was centred on both Shiva and Vishnu (in all his forms), the Northern devotional movement was more or less centred on Rama and Krishna, both of whom were incarnations of Vishnu. Though this did not mean that the cult of Shiva or of the Devi went into decline. In fact for all of its history the Bhakti movement coexisted peacefully with the other movements in Hinduism. It was initially considered unorthodox as it rebelled against caste distinctions and made disregarded Brahmanic rituals which according to Bhakti saints not necessary for salvation. In the course of time however, owing to its immense popularity among the masses (and even royal patronage) it became 'orthodox' and continues to be one of the most important modes of religious expression in modern India.

In the period between the 14-17th centuries, a great bhakti movement swept through Northern India initiated by a looseiy associated group of teachers or 'Sants'. Chaitanya, Vallabh, Meera Bai, Kabir, Tulsi Das, Tukaram and other mystics spearheaded the Bhakti movement in the North. Their teachings were that people could cast aside the heavy burdens of ritual and caste and the subtle complexities of philosophy and simply express their overwhelming love for God. This period was also characterised by a spate of devotional literature in vernacular prose and poetry in the ethnic languages of the various Indian states or provinces.

As aforementioned whilst many of the Bhakti mystics focused their attention on Krishna or Rama, it did not necessarily mean that the cult of Shiva was marginalised. The growth of the Vira-Shaiva and the older Shaiva Siddhanta schools in this period, which incorporated Bhakti into their teachings are testimony to the growth of the Shaiva faith in this period.

In the thirteenth century Basava founded the Vira-Shaiva school or Virashaivism. He rejected the caste system, denied the supremacy of the Brahmins, condemned ritual sacrifice and insisted on bhakti and the worship of the one God, Shiva. His followers were called Vira-Shaivas, meaning "stalwart Shiva-worshippers".

The Shaiva-Siddhanta school is a form of Shaivism (Shiva worship) found in the south and is of hoary antiquity. It incorporates the teachings of the erstwhile Shaiva nayanars and espouses the belief that Shiva is Brahman and his infinite love is revealed in the divine acts of the creation, preservation and destruction of the universe, and in the liberation of the soul.

Seminal Bhakti works in Bengali include the many songs of Ramprasad Sen. His pieces (known as Shyama Sangeet, or Songs of the Dark Mother) are still actively sung today in West

Bengal. Coming from the 17th century, they cover an astonishing range of emotional responses to Ma Kali, detailing complex philosophical statements based on Vedanta teachings and more visceral pronouncements of his love of Devi. Using inventive allegory, Ramprasad had 'dialogues' with the Mother Goddess through his poetry, at times chiding her, adoring her, celebrating her as the Divine Mother, reckless consort of Shiva and capricious Shakti, the universal female creative energy, of the cosmos.

Rama Bhakti

The leader of the bhakti movement focusing on the Lord as Rama was Ramananda. Very little is known about him, but he is believed to have lived in the first half of the 15th century. He taught that Lord Rama is the supreme Lord, and that salvation could be attained only through love for and devotion to him, and through the repetition of his sacred name.

Ramananda's ashram in Varanasi became a powerful centre of religious influence, from which his ideas spread far and wide among all classes of Indians. One of the reasons for his great popularity was that he renounced Sanskrit and used the language of the people for the composition of his hymns. This paved the way for the modern tendency in northern India to write literary texts in local languages.

Devotees of Krishna worship Him in different mellows, known as rasas. Two major systems of Krishna worship developed, each with its own philosophical system. These two moods as called aishwaryamaya bhakti and madhuryamaya bhakti. Aishwaryamaya bhakti is revealed in the abode of queens and kingdom of Krishna in Dwarka. Madhuryamaya Bhakti is revealed in the abode of braja.

Thus Krishna is variously worshipped according to the development of devotee's taste in worshipping the Supreme Personality of Godhead, Sri Krishna, as father, friend, master, beloved and many different varieties which are all extraordinary. Krishna is famous as Makhanchor, or butterthief. He loved to eat butter and is the beloved of his little village in Gokul. These are all transcendental descriptions. Thus they are revealed to the sincere devotees in proportion to the development in their love of Godhead.

Shri Madhvacharya (1238-1317) identified God with Vishnu. His view of reality is purely dualistic in that he understood a fundamental differentiation between the ultimate Godhead and the individual soul, and the system is therefore called Dvaita (dualistic) Vedanta. Madhva is considered one of the influential theologians in Hindu history. His influence was profound, and he is one of the fathers of the Vaishnava Bhakti movement. Great leaders of the Vaishnava Bhakti movement in Karnataka like Purandara Dasa, Kanaka Dasa, Raghavendra Swami and many others were influenced by Dvaita traditions.

Vallabhacharya (1479-1531) called his system of thought Shuddhadvaita (pure monism). According to him, it is by God's grace alone that one can obtain release from bondage and attain Krishna's heaven. This heaven is far above the "heavens" of Brahma, Vishnu and Shiva, for Krishna is himself the eternal Brahman.

Chaitanya Mahaprabhu (1486-1534) defined his system of philosophy as Achintya Bheda Abheda (inconceivable and simultaneous oneness and difference). It synthesizes elements of monism and dualism into a single system. Chaitanya's philosophy is taught by the contemporary International Society for Krishna Consciousness, better known as the *Hare Krishna* movement. Srimanta Sankardeva (1449-1568) named his religion *ek sarana naam dharma* and propagated it in Assam. An example of *dasa bhakti*, in this form there was no place for Radha. The most important symbol of this religion is the *naamghor* or prayer hall, which dot Assam's landscape. This form of worship is very strong in Assam today, and much of the traditions are maintained by the monasteries called *Satras*.

Vaishnava Bhakti

For more information about leaders of the Vaishnava bhakti movement, see Vaishnavism, Gaudiya Vaishnavism and Hare Krishna.

Prominent historical personalities include:

- Ramanuja

- Nimbarka
- Madhva
- Vallabh
- Chaitanya
- Bhakti Tirtha Swami

Influences

Beyond the confines of such formal schools and movements, however, the development of bhakti as a major form of Ḥindu practice has left an indelible stamp on the faith. Philosophical speculation was concern for the minority, and even the great Advaitist scholar Adi Shankaracharya, when questioned as to the way to God, said that chanting the name of the lord, was essential. The philosophical schools changed the way people thought, but Bhakti was immediately accessible to all, calling to the instinct emotion of love and redirecting it to the highest pursuit of God and self-realization.

In general a liberal movement, its denouncement of caste offered recourse for Hindus from the orthodox Brahaminical systems. Of course, however, Bhakti's message of tolerance and love was not often heeded by those ensconced in the societal construct of caste. Altogether, Bhakti resulted in a mass of devotional literature, music and art that has enriched the world and gave India renewed spiritual impetus, one eschewing unnecessary ritual and artificial social boundaries.

Shaktism

Shiva and Shakti as One

Shaktism is a denomination of Hinduism that worships Shakti, or Devi Mata—the Hindu name for the Great Divine Mother—in all of her forms whilst not rejecting the importance of masculine and neuter divinity (which are however deemed to be inactive in the absence of the Shakti). In pure Shaktism, the Great Goddess, or Mahadevi, is worshipped as nothing less than the highest divinity, Supreme Brahman Itself, the "one without a second," with all other forms of Divinity, female or male, considered to be merely her diverse manifestations.

Philosophy

In his seminal *History of the Shakta Religion*, N.N. Bhattacharyya explained that "(those) who worship the Supreme Deity exclusively as a Female Principle are called Shakta. The Shaktas conceive their Great Goddess as the personification of primordial energy and the source of all divine and cosmic evolution. She is identified with the Supreme Being, conceived as the Source and the Spring as well as the Controller of all the forces and potentialities of Nature. Nowhere in the religious history of the world do we come across such a completely female-oriented system."

Alternative interpretations of Shaktism, however—primarily those of Shaivite scholars, such as Satguru Shivaya Subramuniyaswami—argue that the feminine manifest is ultimately only the vehicle through which the masculine Unmanifest Parashiva is ultimately reached. In this interpretation, the Divine Mother becomes something of a mediatrix, who bestows advaitic moksha on those who worship Her. Thus, these Shaivite views often conclude that Shaktism is effectively a sub-denomination of Shaivism, arguing that Devi is worshipped in order to attain union with Shiva, who in Shaktism is the impersonal unmanifest Absolute. This remains a minority view in Shaktism proper, which considers Shiva as an equal and inseparable aspect of Devi.

Origin and History

Shaktism as we know it today developed between the 4th and the 7th centuries CE in India. It was during this development that the many religious texts, known as the Tantras, were written. In a certain sense, one could consider oneself a Shakta (a devotee of Shakti), a Shaiva (a devotee of Shiva), and a Vaishnava (a devotee of Vishnu) all at the same time.

Roots in Hinduism

This form of Hinduism is strongly associated with Vedanta, Samkhya and Tantra Hindu philosophies and is ultimately monist, though there is a rich tradition of Bhakti yoga associated with it. The feminine energy (Shakti) is considered to be the motive force behind all action and existence in the phenomenal cosmos in Hinduism. The cosmos itself is Brahman, the concept of the unchanging, infinite, immanent and transcendent reality that is the Divine Ground of all being, the "world soul". Masculine potentiality is actualized by feminine dynamism, embodied in multitudinous goddesses who are ultimately reconciled in one.

The keystone text is the Devi Mahatmya which combines earlier Vedic theologies, emergent Upanishadic philosophies and developing tantric cultures in a laudatory exegesis of Shakti religion. Demons of ego, ignorance and desire bind the soul in maya (illusion) (also alternately ethereal or embodied) and it is Mother Maya, shakti, herself, who can free the bonded individual. The immanent Mother, Devi, is for this reason focused on with intensity, love, and self-dissolving concentration in an effort to focus the *shakta* (as a Shakti worshipper is sometimes known) on the true reality underlying time, space and causation, thus freeing one from karmic cyclism. A common hymn describing the 1000 names of Devi is the Lalita sahasranama.

Worship

Among the manifestations of Devi most favoured for worship by Shaktas are Kali, Durga, and Parvati. Durga is an epithet of Mahadevi, or "Great Goddess," who is celebrated in the Devi Mahatmya. Kali is the goddess of destruction and transformation, as well as the devourer of time, as her name implies (*kala* means "time," and also means "black"). Parvati is the gentle wife of Shiva, one of the most popular gods of modern Hinduism, and is strongly associated with Kali and other goddesses.

Shakta worship takes many forms, but is heavily influenced by Tantra. Shakti is worshipped in several ways in the course of a puja (worship ceremony), including offerings of sweets and flowers, chanting mantras, using mudras, and typically offering some sort of sacrifice. She is most powerfully worshipped by chanting her bija mantra, which is different for each goddess.

Animal sacrifice is performed in some places in India, including such major sites as Kalighat in Calcutta, West Bengal, where goats are sacrificed on days of Tuesdays and Saturdays, and Kamakhya in Guwahati, Assam. Black male goats are typically sacrificed, as well as male buffalo during Durga Puja, and this practice is a controversial one. The brahmin performing the sacrifice is not allowed to cause pain to the animal, and must wait for the animal to surrender before cutting off the head with a single stroke. The blood is used to bless icons and worshippers, and the meat cooked and served to the worshippers and poor as prasad. Those who are averse to animal sacrifice will use a pumpkin or melon instead, which has become a popular and acceptable substitute.

Shaktism is also fused with local beliefs in villages throughout India. In Tamil Nadu, Kerala and Andhra Pradesh, she is known as Amma (mother). Rural Bengalis know her as Tushu. The Brahmanical idea of Shakti has become fused with local beliefs in protective village goddesses who punish evil, cure diseases and bring boons and blessings to the people of the village. Major annual festivals throughout India include Durga Puja (October, national), Diwali (November, national), Kali Puja (October/November, national), Minakshi Kalyanam (April/May in Madurai, Tamil Nadu) and Ambubachi Mela (June/July in Kamakhya Temple, Guwahati, Assam), which is the most important festival to Shakta Tantriks.

Shakta Upanishads

Of the 108 Upanishads of the Muktika, 9 are considered Shakta Upanishads. They are listed with their associated Veda (ZV, KYV, AV):

1. S+t (AV)
2. Annapkrga (AV)
3. Dev+ (AV)
4. Tripurtapani (AV)
5. Tripura (RV)
6. Bhvana (AV)
7. Saubhgya (RV)
8. Sarasvat+rahasya (KYV)
9. Bahvsca (RV)

Shakti in Ritual

"The goddess is the Adya Shakti, the original energy out of which all things were created. This form has many aspects, and all of them have to be honoured and served in the remaining days of the Durgapuja. The long series of oblations, libations, and flower offerings are devoted to the recognition of and deference to the multiple aspects in the original form of the goddess. This recognition is central to shakti puja; *bhed*, division within the one goddess, must be revealed so that the full significance of Durga may be comprehended and everything may be reintegrated into the idea, form, and appearance of the goddess.

These aspects must be adored separately; otherwise the goddess is not satisfied."

Source : "The Play of the Gods: Locality, Ideology, Structure and Time in the Festivals of a Bengali Town", Akos Ostor, University of Chicago Press (1980).

Shiva

What is the difference between the soul and the mind?

The soul is the consciousness or the knowing principle in man.

The mind belongs to the category of matter.

Shiva is the soul and Parvati is the mind.

The consciousness is the Reality, the unchanging,

unmoving, all pervading soul.

Mind being matter is changeable and unreal.

We perceive objects through the organs of sense perception. The soul enables the organs to perceive the objects. The soul exists in all living beings as consciousness. It is the light of the soul that makes the sense-organs and the mind appear alive and luminous.

Shiva is the unchanging consciousness – Nirguna or without form or shape. This unchanging consciousness-Shiva – becomes saguna or with form, when Maya Shakti, which is Shiva's illusive power appears as mind and matter. This phenomenon is known as Ardhanarishwar.

Shiva's Maya-Shakti is known by various names such as Uma, Parvati, Kali, Durga etc. This Maya-Shakti or power inheres in Shiva just as the burning power inheres in fire, sweetness in sugar, whiteness in milk and meaning in the words.

Shiva stands for the Absolute, the unchanging, static background, of which Kali, the Shakti (power) is the dynamic expression. We call this Shakti or power Mother or Goddess. Goddess Kali combines in herself creative dynamism, destructive terror and redemptive grace.

God's power that is enveloped by Tamo-guna-pradhana is Lord Shiva. Shiva's trishul or trident (the three pointed javelin like weapon) represents the three gunas –Sattwa, Rajas and Tamas. The Trishul is the emblem of sovereignity. Lord Shiva wields the world through these three gunas, which is the composition of Maya-Shakti or mind and matter. In other words, if one were able to analyse the composition of Maya-Shakti, it will be found to be made of the combination of the three Gunas (Sattwa-Rajas-Tamas).

Shiva is Trilochana, the three-eyed one, in the centre of whose forehead is the third eye, the eye of wisdom (gnana-chakshu). The burning power of the wisdom of the third eye destroys desires for worldly objects. Sensual desire and lust is represented by Kamadev, the god of love (Eros or Cupid). When a person reaches a state of perfect renunciation-he is said to have burnt all his desires. Shiva's third eye burnt to ashes Kamadev-the god of love. The eye of wisdom leads to transcendental vision of the Supreme Reality.

We recite regularly the following Maha Mrityunjay Mantra from the Sukla Yajurveda Samhita III.

Om Trayambakam Yajaamahe
Suganghim Pushtivardhanam
Urvaarukmiva Bandhanaan
Mrityor Mokshiya Mamritaat

(The meaning of the Mantra is as follows)

I worship thee, O sweet Lord of transcendental vision (the three-eyed one or Lord Shiva). O giver of prosperity to all, may I be free from the bonds of death, even as a melon (or cucumber) is severed from its bondage or attachment to the creeper.

The word Shiva signifies the auspicious. The good Lord Shiva roots out sin and terror, and is the bestower of earthly happiness, promoter of good and auspiciousness. Shiva is also called Samkara which means doer of good. Shiva takes one beyond the three bodies (Tripura), gross, subtle and causal that envelope the Jiva or the embodied soul. He is hence the Hara or the remover of all evil and the ideal of renunciation. Therefore we hail 'Hara hara Mahadeva.'

In the Rigveda (x,121,4) it is writtena Yasya ime himavanto mahitva. That the snow capped Himalayas appear as if they are meditating (dhyayativa), and they are the concrete symbol of the glory of the Supreme. Hence it is no wonder that most of its attributes are transferred to Shiva. Kailash (mountain) in the Himalayas is the abode of Lord Shiva.

The snowy Himalayas are white and Shiva's body is also white – Gauranga (gauri = light complexioned). Karpura gaura = camphor hued white. Shiva's body is smeared with bhasma or ashes to indicate renunciation, whitenes and purity.

Amazing! Lord Shiva is present in the Kailas mountain. The most sacred symbol in Hinduism AUM imprinted with snow and ice on the face of the mountain by nature. Snow falls each winter and resides on the mountain top in the shape of Om. After Arati (devotional adoration of the Lord with waving of lamps) we usually recite the following prayera

Karpur gauram karunaa avataaram,
sansaar saaram Bhujgendra haaram,
Sadaa vasantam hridayaarvinde,
Bhavam Bhavaani sahitam namaami
I bow to that camphor-hued, white complexioned
(Lord Shiva), who is Incarnation of compassion,
who is the very essence of (consciousness; the
knowing principle) of life (of the embodied soul);
Who wears snakes as garlands, whose eternal abode

is in the heart of the devotee, I bow to Him (Lord Shiva) and His consort Bhavani (Uma or Paarvati).
Karpur (camphor-hued); Gauram (white);
Karunaa (compassion); Avataaram (incarnation); Sansaar
(life of the embodied soul); Saaram (essence, the knowing principle or consciousness);
Bhujagendra (wearer of snakes or who wields the Serpent power of Kundalini Shakti); Haaram (garlands); Sadaa (eternal); Vasantam (resides);
Hridayaarvinde (in the heart of the devotee);
Bhavam (Lord Shiva); Bhavaani (Uma or Paarvati);
Sahitam (together); Namaami (I bow).

The rain water is locked up in the Himalayas as snow or ice and river Ganga (Ganges) falling from the heaven is locked up in the Jata (matted locks) of Shiva. Hence Shiva is called Ganga-dhara. River Ganga issues from the Himalayas, and Ganga flows down to earth from Shiva's matted locks after release. The holy river flows down from Shiva's head and therefore it symbolises the stream of wisdom.

Snakes are symbolic of the mental powers (the coiled up serpent power of Kundalini Shakti) under the control of the divinity. The moon symbolises mind in a state of tranquility and purity. The Damaru (a small drum-like instrument in Lord Shiva's hand) represents the sabda Brahman. This is AUM and the sound of AUM (OM), from which all languages are formed.

Rudraksha Beads

The Mala (rosary) is made from the Rudraksha beads. The Shiva Purana (25 chapter) describes Lord Shiva, the Yogeshwar (master of Yoga), meditating for thousands of years for the benefit of people everywhere. According to the legend, when Shiva opened his eyes, some tear-drops fell on earth and grew into Rudraksha trees. These trees grow in several parts of India. Both the Shiva Mahapurana and the Devi Bhagavatam describe Rudraksha beads as highly auspicious. The mere looking at Rudraksha beads creates auspiciousness. Touching the Rudraksha beads multiplies the auspiciousness manifold, and the wearing of Rudraksha Mala (rosary) augurs almost continuous flow of auspiciousness.

Rudraksha beads are found with from one eye to up to fourteen eyes or fourteen sided beads. Their medicinal and other uses are specific to how many sides there are on the beads. For examplea: One sided Rudraksha bead is producer of worldly happiness and liberation, producer of wealth, destroyer of obstacles and problems, fulfiller of wishes and highly effective in tuberculosis type of diseases.

Two sided Rudraksha bead helps to increase mental powers, calms agitated minds, helps to overcome Tamasic Guna.

Three sided Rudraksha bead helps in acquiring knowledge and skills, helps increase digestive power, effective in reducing fever and in eye diseases.

Four sided Rudraksha bead works wonders in increasing memory, especially helpful to those with weak memory. Also improves power of speech. The procedure is to drink for twenty days, milk boiled with Rudraksha beads.

Since Shiva is the unchanging consciousness-Nirguna or without form, how to give a form to the formless for the purpose of worship? This dilemma is solved through the symbol of the Shiva-Linga. Like the inverted bowl with the limitless rim called the sky, the Shiva-Linga represents visible infinity. When Shiva and Shakti are separated into a duality of chit and sat-consciousness and manifest existence or matter (subject and object), the universe of different planes of existence comes into being. This is variously described as spirit and matter, Purusha and prakriti, Brahman and Maya, Shiva and Shakti, Linga and Yoni etc.

A Legend

When creation was completed, Shiva and Parvati went to live on top of Mount Kailash.

Parvati askeda "O adorable Lord, which of the many rituals observed in your honour does please you most?"

Lord Shiva replieda "The fourteenth night of the new moon in the dark fortnight during the month of Phalgun is my favourite day. It is called 'Shivaratri'. My devotees give me greater happiness by mere fasting than by ceremonial baths and offerings of flowers, sweets and incense. The offering of a few bilwa (bael) leaves is more precious to me than precious jewels and flowers. My devotee must observe strict spiritual discipline during the day and worship me at night".

Parvati was deeply impressed by the words of Lord Shiva. She repeated them to her friends, who in their turn passed them on to the ruling princes on earth. Thus was the sanctity of Shivaratri broadcast all over the world.

On Mahashivaratri, the devotees observe strict spiritual disciplines during the day. The devotees worship Lord Shiva at night in four different ways during each of the four successive three hour periods of the night. Bathing the Shiva-Linga in milk in the first period, in curd during the second period, in ghee in the third period and in honey in the fourth and last period. Every three hours, a round of worship of Shiva Linga is conducted. Evils like lust, anger, jealousy, born of Rajas and Tamas, are subdued. The devotees break the fast after the fourth and last round. In the morning, the devotees should feed the Brahmins (priests) first and after doing the prescribed ceremonies, break the fast. The devotees observe vigil throughout the night. After the completion of the rites of Shivaratri, the devotee presents gifts or donations to the officiating priests.

According to Vedic scriptures, the performance of this ritual is both obligatory and desirable. The injunction laid down for the performance of the ritual for transforming the devotee's body into a residence fit for the divinity area non-injury to living creatures, truthfulness, freedom from anger, celibacy, compassion, forbearance, austerities, calmness, freedom from passion and malice. As a reward, it is said that one who performs the sacrifice of Shivaratri with all the attendant rituals and keeps the fast according to the rules laid down gets happiness and realises his most cherished desires.

In the case of Shiva as chief of ascetics, no food is generally offered as prasad. The daily ceremonials are of austerely simple kind. Water is poured on Shiva-Linga with perhaps a few oblations of flowers and bilwa leaves. It is remarkable that even in cases where food is offered to this divinity it is not allowed to be eaten by his votaries. According to the Brahminical rule, 'leaves, flowers, fruits and water become unfit to be consumed after being consecrated to Shiva'.

The three most important religious activities during the Maha-Shivaratri festival area fasting during the entire lunar day, keeping a strict night vigil and worshipping the lingam with offerings of foods, leaves, flowers etc. and with recitation of mantras. The two great natural forces that afflict man are Rajas (quality of passion) and Tamas (inertia). The Shivaratri Vrata aims at the perfect control of these two qualities.

The motivation behind the vrata or fast is mainly the promotion of physical and mental self-control by acts of penance, thanks giving, praise of the Supreme Lord and for desiring special boon.

Swami Shivananda of the Divine Life Society, Rishikesh, advised to offer this inner worship to Lord Shiva dailya "O Lord Shiva, Thou art my self. My mind is Parvati. My pranas are Thy servants. My body is Thy house. All my actions in this world are Thy worship. My sleep is samadhi. My walk is circumambulation around Thee. My speech is prayer unto Thee. Thus do I offer unto Thee all that I am." The following is from the Tulasi Ramayana, Uttara-Kanda, Doha 45a "Shankara Bhajan Bina Nara Bhagati Na Paavai Mori" Sri Rama saida "With joined palms I lay before you all anothersecret doctrinea without adoring Shankara (Lord Shiva) man cannot attain devotion to Me."

Lord Shiva

Lord Shiva has at least four quite distinct characters, each of which has a female or active energising counterpart (shakti).

In the first place, as Shiva, Sada-Shiva, Shankara or Shambhu, the eternally blessed one or the source of blessings. He is the eternal reproducing power of nature, perpetually restoring and reproducing itself after dissolution, under which Shiva is often identified with the eternal creative essence, the great eternal Supreme being as Maha-Deva or the Supreme

Lord termed Ishwara. Hence in this aspect Shiva is represented by the symbol of the Lingam and the Yoni combined.

Temples that hold this Shiva's emblem or symbol, which is of a double form to express the blending of the male and female principles in creation, are probably the most numerous of any temples to be seen in India. There can be no doubt, in fact, that the Supreme creative power is universally worshipped throughout India, under the name of Shiva and of his consort Jagan-Matri, or Mother of the universe.

In the second place, as Maha-Yogi is the great representative Yogi or Tapasvi, who has attained the highest perfection and meditation and austerity. In this aspect Shiva appears as an austere naked ascetic (Digambara) with body covered with ashes and matted hair (dhurjati), abiding fixed and immovable in one spot (sthanu), teaching men by his own example the power to be acquired by renunciation, suppression of passions, and abstract contemplation and meditation as leading to the highest spiritual knowledge. In this aspect of the yogi, as in that of the reproducer, He is also sometimes called the 'Blessed one' (Shiva).

In the third place, Shiva is the entire reverse of the ascetical. In this aspect, living in the Himalaya mountains with his consort Parvati, often dancing with her the Tandava dance, He is surrounded by dwarfish troops (gana). This is the aspect in which He is worshipped by Tantrikas. In the fourth place, as Rudra or Mahakala, he is the destroying or dissolving power of nature; when he is either a personification of all matter resolving itself into its constituent elements or of Kala (Time), the great dissolver. The more active principle of destruction being assigned to his consort Kali.

In the fourth place, there are yet two other aspects of Shiva. In the first of these, as the dissolver of the universe, He is the terrible destroyer (Bhairav), with His consort Kali engaged in the active role.

In the second of these, He is also called Bhuteswara, Lord of spirits or demons, haunting cemeteries and burial grounds, wearing serpents for garlands, and a string of skulls for a necklace. As Bhuteswara, He is sometimes surrounded by troops of imps and spirits (bhuta), and sometimes He is trampling on rebellious demons who have acquired too great power.

Here we may observe that in every one of his aspects, the consort of Shiva is not only His counterpart, but generally represents an intensification of his attributes. As destructress, She is Kali, as reproducer she is symbolised by the Yoni (Shiva's emblem). She is the mother of the universe (Jagan-matri). She is the type of beauty in Uma. She has also her forms as a female ascetic (yogini). In her role as destructress, she is Bhairavi Durga. As a mountaneer, she is Parvati. All these attributes are combined in her aspect of Mother Durga.

Shiva and Parvati represent the gathering together, and unifying in one personality, numerous attributes, properties and functions belonging to various deities and various divine forces. The destructive energies of the atmosphere exhibited in wind and storm and personified in the Vedas as Vayu, Rudra and the Maruts; the all consuming potency of time; the fertilising properties present in dew and rain; the almighty agencies operating in creation and the same agencies operating as re-creation or reproduction; the power of asceticism exhibited in the Maha-yogi; the terrific frightful agencies and operations of demons and spirits; as Shiva, Sada-Shiva, Shankara, Shambhu-the eternally blessed one or causer of blessings; He is the eternal reproducing power of nature, perpetually restoring and reproducing itself after dissolution. Under which He is identified with the eternal creative essence, the great eternal Supreme Being as Maha-Deva or Supreme Lord Ishwara.

Hence in this aspect, Shiva is represented by the symbol of the linga and yoni, rather than by any human personification. The Shiva-Linga symbolises the blending of the male and female principles in creation. This Supreme creative power is universally worshipped throughout India under the name of Shiva and His consort Jagan-Matri-mother of the universe-all these have been centralised in Shiva and His consort personified as half male and half female known as Ardhanarishwara (symbolising the union of spirit and matter).

Aswatthaman, the son of Drona, saida I shall at this hour seek the protection of the puissant Mahadeva! I will take the shelter of that god, that source of everything beneficial, *viz.*, the lord of Uma, otherwise called Kapardin, decked with a garland of human skulls, that plucker of Bhaga's eyes called also Rudra and Hara. In ascetic austerities and prowess, he far surpasses all the gods. I shall, therefore, seek the protection of Girisha (Shiva) armed with the trident.

I seek the protection of Him called Fierce, Stanu, Shiva, Rudra, Sarva, Isana, Iswara, Girisha; and of that boon giving god who is the Creator and Lord of the universe; of Him whose throat is blue, who is without birth, who is called Sakra, who destroyed the sacrifice of Daksha, and who is called Hara; of Him whose form is the universe, who has three eyes, who is possessed of multifarious forms, and who is the lord of Uma; of Him who resides in crematoriums, who swells with energy, who is the lord of diverse tribes of ghostly beings, and who is the possessor of undecaying prosperity and power; of Him who wields the skull-topped club, who is called Rudra, who bears matted locks on his head, and who is a Brahmacharin (celibate). Purifying my soul that is so difficult to purify, and possessed as I am of small energy, I adore the Destroyer of the triple city, and offer myself as the victim. Hymned thou hast been, deserving art thou of hymns, and I hymn to thy glory!

Thou art robed in skins; thou hast red hair on thy head. Thou art pure; thou art the Creator of Brahman; thou art Brahma; thou art an observer of vows; thou art devoted to ascetic austerities; thou art the refuge of all ascetics; thou art the leader of diverse tribes of ghostly beings; thou art three eyed; thou art dear to Gauri's heart; thou hast for thy excellent bearer a bovine bull; thou art the protector of all quarters; thou hast the moon as an ornament on thy brow!

The Four Faces of Shiva

Uma saida O holy one, why are those faces of thine which are on the east, the north, and the west, so handsome and so agreeable to look at like the very moon? And why is that face of thine which is on the south so terrible? Why are thy matted locks tawny and so erect? Why is thy throat blue after the manner of the peacock's plumes? Why, O illustrious deity, is the Pinaka always in thy hand? Why art thou always a Brahmacharin with matted locks? O lord, it behoveth thee to explain all these to me. I am thy spouse who seeks to follow the same duties with thee. Further, I am thy devoted worshipper, O deity, having the bull for thy mark!

The blessed and holy one (Shiva) saida .... I became four-faced through Yoga-puissance. Thus I showed my high Yoga-power in becoming four faced. With that face of mine which is turned towards the east, I exercise the sovereignty of the universe. With that face of mine which is turned towards the north, I sport with thee, O thou of faultless features! That face of mine which is turned towards the west is agreeable and auspicious. With it I ordain the happiness of all creatures. That face of mine which is turned towards the south is terrible. With it I destroy all creatures.

I live as a Brahmacharin with matted locks on my head, impelled by the desire of doing good to all creatures. The bow Pinaka is always in my hand for accomplishing the purposes of the deities. In days of yore, Indra (king of heaven), desirous of acquiring my prosperity, had hurled his thunderbolt at me. With that weapon my throat was scorched. For this reason I have become blue-throated.

Uma saida O holy one, O lord of all creatures, O foremost of all observers of duties and religious rites, I have great doubt, O wielder of Pinaka, O giver of boons. These ascetics, O puissant lord, have undergone diverse kinds of austerities. In the world are seen ascetics wandering everywhere under diverse forms and clad in diverse kinds of attire. For benefiting this large assemblage of Rishis, as also myself, do thou kindly resolve, O chastiser of all foes, this doubt of mine. What indications has Religion or Duty been said to possess? How, indeed, do men become unacquainted with the details of Religion or Duty to succeed in observing them? O puissant lord, O thou that art conversant with Religion, do thou tell me this.

Maheswara saida Abstention from injury, truthfulness of speech, compassion towards all beings, tranquillity of soul, and the making of gifts to the best of one's power, are the foremost duties of the householder.

Abstention from sexual congress with the spouses of other men, protection of the wealth and the woman committed to one's charge, unwillingness to appropriate what is not given to one, and avoidance of honey and meat, these are the five chief duties. Indeed, Religion or Duty has many branches all of which are fraught with happiness. Even these are the duties which these embodied creatures who regard duty as superior should observe and practise. Even these are the sources of merit.

Bhishma saida Salutation to thee in thy form of vastness! Thou hadst assumed the form of a recluse with matted locks on head, staff in hand, a long stomach, and having thy begging bowl for thy quiver. Salutation to thee in thy form of Brahmacharin. Thou bearest the trident, thou art the lord of the celestials, thou hast three eyes, and thou art high-souled. Thy body is always besmeared with ashes, and thy phallic emblem is always turned upwards. Salutations to thee in thy form of Rudra! The half-moon forms the ornament of thy forehead. Thou hast snakes for the holy thread circling thy neck. Thou art armed with Pinaka and trident. Salutation to thy form of Fierceness. Thou art the soul of all creatures. Thou art the Creator and destroyer of all creatures. Thou art without wrath, without enmity, without affection. Salutation to thee in thy form of Peace!

Shiva and Rudra

Translated by Sri Kisari Mohan Ganguli Shiva has two forms. One of these is terrible, and the other mild and auspicious.

Vasudeva saida O mighty armed Yudhishthira, listen to me as I recite to thee the many names of Rudra as also the high blessedness of that high-souled one. The Rishis describe Mahadeva as Agni, and Sthanu, and Maheswara; as one-eyed, and three eyed, of universal form, and Shiva or highly auspicious.

Brahmanas conversant with the Vedas say that Shiva has two forms. One of these is terrible, and the other mild and auspicious. Those two forms again, are subdivided into many forms. That form which is fierce and terrible is regarded as identical with Agni and Lightning and Surya (fire, lightning and sun). The other form which is mild and auspicious is identical with Righteousness and Water and Chandramas (moon). Then again, it is said that half his body is fire and half is Soma (or the moon). That form of his which is mild and auspicious is said to be engaged in the practice of the Brahmacharya (Celibacy) vow. The other form of his which is supremely terrible is engaged in all operations of destructions in the universe.

Because he is great (Mahat) and the Supreme Lord of all (Iswara), therefore he is called Maheswara. And since he burns and oppresses, is keen and fierce, and endued with great energy, and is engaged in eating flesh and blood and marow, he is said to be Rudra. Since he is the foremost of all the deities, and since his dominion and acquisitions are very extensive, and since he protects the extensive universe, therefore he is called Mahadeva. Since he is of the form or colour of smoke, therefore he is called Dhurjati. Since by all his acts he performs sacrifices for all and seeks the good of every creature, therefore he is called Shiva or the auspicious one.

Staying above (in the sky) he burns the lives of all creatures and is, besides, fixed in a particular route from which he does not deviate. His emblem, again, is fixed and immovable for all time. He is, for these reasons, called Sthanu. He is also of multiform aspect. He is present, past and future. He is mobile and immobile. For this he is called Vahurupa (of multiform aspect). The deities called Viswedevas reside in his body. He is, for this, called Viswarupa (of universal form). He is thousand-eyed; or he is myriad-eyed; or, he has eyes on all sides and on every part of his body. His energy issues through his eyes. There is no end of his eyes. Since he nourishes all creatures and sports also with them, and since he is their lord and master, therefore he is called Pasupati (the lord of all creatures).

If there is one who worships him by creating his image, another who worships his emblem, the latter it is that attains to great prosperity for ever.

Since his emblem is always observant of the vow of Brahmacharya, all the wolds worship it accordingly. This act of worship is said to gratify him highly. If there is one who worships him by creating his image, another who worships his emblem, the latter it is that attains to great prosperity for ever. The Rishis, the deities, the Gandharvas, and the Apsaras, worship that emblem of his which is ever erect and upraised. If his emblem is worshipped, Mahadeva becomes highly gratified with the worshipper. Affectionate towards his devotees, he bestows happiness upon them with a cheerful soul.

This great god loves to reside in crematoria and there he burns and consumes all corpses. Those persons that perform sacrifices on such grounds attain at the end to those regions which have been set apart for heroes. Employed in his legitimate function, he it is that is regarded as the Death that resides in the bodies of all creatures. He is again, those breaths called Prana and Apana in the bodies of all embodied beings. He has many blazing and terrible forms. All those forms are worshipped in the world and are known as Brahmanas possessed of knowledge. Amongst the gods he has many names all of which are fraught with grave import. Verily, the meanings of those names are derived from either his greatness or vastness, or his feats, or his conduct. The Brahmanas always recite the excellent Sata-rudriya in his honour, that occurs in the Vedas as also that which has been composed by Vyasa. Verily, the Brahmanas and Rishis call him the eldest of all beings. He is the first of all the deities, and it was from his mouth that he created Agni.

That righteous-souled deity, ever willing to grant protection to all, never gives up his suppliants. He would much rather abandon his own life-breaths and incur all possible afflictions himself. Long life, health and freedom from disease, affluence, wealth, diverse kinds of pleasures and enjoyments, are conferred by him, and it is he also who snatches them away. The lordship and affluence that one sees in Sakra and the other deities are, verily his. It is he who is always engaged in all that is good and evil in the three worlds.

In consequence of his fullest control over all objects of enjoyment he is called Iswara (the Supreme Lord or Master).

Since, again, he is the master of the vast universe, he is called Maheswara. The whole universe is pervaded by him in diverse forms. It is that deity whose mouth roars and burns the waters of the sea in the form of the huge mare's head. (Note: The allusion is to the fiery mare's head which is supposed to wander through the ocean.)

Mahashivaratri (Magha Krishna Chaturdasi)

(The 14th day of the dark half of every month-Krishna Chaturdasi is called 'Shivaratri' or 'Maha-Shivaratri'. The one in the month of Magha (February-March) is christened 'Mahashivaratri', since it is the greatest of all.)

Of all the major Hindu festivals, Mahashivaratri is the only one wherein the austerity part (as signified by the very word 'vrata') is predominant. There is practically no festivity, revelry or gaiety in its observance, the whole thing being one of continuous solemnity. This is but natural since Shiva is the god of the ascetics, the very incarnation of vairagya or renunciation! This vrata is open to all human beings. The basic disciplines to be kept up on this day are ahimsa (non-injury), satya (speaking the truth), Brahmacharya (continence), daya (compassion), Ksama (forgiveness) and anasuyata (absence of jealousy).

Fasting is one of the most essential aspects of this vrata. So also jagarana or keeping vigil in the night. Worship of Shiva throughout the night, bathing the Shivalinga with panchamrta (five tasty things-milk, curds, ghee, sugar and honey), homa, japa of the mulamantra (basic mantra, *viz.*, Om Namas Shivaya) and prayer for forgiveness-are the other items involved in its observance. Abhisheka is a part of the worship of Lord Shiva. Without it, the worship is incomplete. It is the ceremonial bathing of the Shiva Lingam in Shiva temples. A pot made of copper or brass, with a tiny hole in the centre, is kept hanging over the image or Lingam of Shiva. The water drips (falls) on the image throughout the day and night. Pouring water, milk, ghee (clarified butter), curd, honey or coconut water over the Lingam is also Abhisheka. Whilst this is done, the Rudram is chanted loudly with devotion and love. Lord Shiva is invoked by performing this Abhisheka.

Monday is a very important day for worshipping Lord Shiva. The thirteenth day (Pradosha) of the bright and dark fortnights is also considered sacred. On these days, devotees of Lord Shiva offer special worship with plenty of prasad. The water of the Abhisheka is considered very sacred. It is known to grant great benefits on the devotees who take it as the Lord's prasad. It purifies the heart and destroys countless sins. You should take it with intense faith and devotion. When you perform Abhisheka with devotion, your mind is concentrated. Your heart is filled with divine thoughts and with the image of the Lord. You forget your body and your surroundings. Egoism vanishes. When the body is forgotten, you begin to enjoy and taste the eternal bliss of Lord Shiva. The recitation of Mantras during the Abhisheka purifies the mind. The greatest and the highest Abhisheka is to pour the waters of pure love on the Lingam in the lotus of the heart. The external Abhisheka with objects is intended to lead to this internal Abhisheka, wherein there is a flow of pure love.

The sacred prasad of the Lord and the holy water of the Abhisheka purify the heart if taken with faith and devotion. They can bring peace and prosperity. Incurable diseases are cured by performing Abhisheka. It bestows health, wealth, prosperity, peace of mind and purity of heart. It expands the heart. It calls for self-sacrifice and self-surrender. There must be a natural feeling in the heart. "I am Thine, my Lord. All is Thine, my Lord." Of all the Hindu Gods none appealed to the emotions of a Hindu mind as much as Lord Shiva, the dweller of the mountains and the bearer of the trident, who with His third eye symbolizes the all knowing awareness of the very Brahman.Although universally acknowledged as one of the trinity gods of Hinduism, to His followers, He is Maheswara, representing the Trinity all by Himself, His different aspects manifesting themselves as the creator, preserver and destroyer of the worlds He creates. Symbolically He is worshipped as Lingeswara, which name stands more for His creative prowess than sexual.

If Brahma is the creator, Vishnu the preserver, Shiva is the quintessential destroyer. His duty is to destroy all the worlds at the end of creation and dissolve them into nothingness. Modern theories of space do suggest the possible ending of the physical universe after some billions of years through the expansion of a gigantic black hole devouring the matter from endless galaxies. Perhaps Shiva would be the black hole performing this task. However this does not mean that Shiva would remain idle till the arrival of that time. Before the worlds really come to an end, Shiva has many things to do to keep the worlds going. His first and foremost task is to destroy many things in order to ensure the Rta or the order of the universe. Shiva's destruction is not negative. It is a positive, nourishing and constructive destruction that builds and transforms life and energy for the welfare of the world and the beings that inhabit it. He destroys in order to renew and regenerate. His destruction is the destruction of an artist, or a surgeon or a cook. Through destruction he facilitates the smooth transitions of things and events from one stage to another.

He destroys our imperfections in order to ensure our spiritual progress. He destroys our illusions, desires and ignorance. He destroys our evil and negative nature. He destroys our old memories, so that we can move on with the movement of time. He destroys our relationships, attachment, impurities, physical and mental wrong doings, the effects of bad karma, our passions and emotions and many things that stand between us and God as impediments to our progress and inner transformation. And in the end when we have made sufficient progress, when we are ready and prepared, and when we are willing without any inner conflict, he destroys death. Historians believe that He was a pre-Vedic god who was admitted into the Vedic Pantheon because of His immense popularity among many non-Aryan tribes. There is a theory that the seated yogi found on the seals of Indus Valley Civilization, was a prototype of Lord Shiva only. References to Him are found in the Rigveda.

In the Mahabharata we come across several references to Him. Both Arjuna and Krishna worshipped Him and obtained favours from Him. In the Ramayana too we come across several references to Him. Sage Valmiki depicted Ravana, the demon king and chief villain of the epic, as a great devotee and daily worshipper of Lord Shiva. It is believed that worship of His icons was widely prevalent as early as 2nd or 3rd Century B.C.

With the popularity of Shaivism a great deal of literature grew around Him, which came to be recognized as Agama literature. So great was His popularity and fame that Agama literature was placed on equal footing to even the Vedas by His followers. In the Svetavatara Upanishad He was elevated to the status of Brahman, by the sage who composed it, after he had a vision of Lord Shiva as the Absolute and Supreme Brahman. Shaivism as a popular movement took shape mainly in South India because of the patronage of the Pallavas and the dedicated work of many great Tamil Saints, who built enormous bhakti or devotional literature in His honour and made His name a household name.

Megasthenes mentioned the worship of Shiva in his book Indika. So did Patanjali, the composer of Yoga Sutras. A great devotee of Shiva named Lakulisa lived some time during the early or pre Christian era and founded the Pasupatha sect, making in the process, Shaivism a very popular movement in the sub-continent. Two great rulers of ancient India, Kanishka and Harshavardhana were His great followers. In the south the Cholas built many temples in His honour. He is known by many names and tittles. As a god of anger, He is Rudra, the Red One, feared by one and all. As Kailasapathi, He is the Lord of Kailash, His Abode in the Himalayas. As Purusha, He is Iswara Himself. As the Lord of the beings, He is known as Pasupathinath. As the husband of Uma, the Mother Goddess, He is known as Umapathi or Parvathipathi.

As the bearer of the sacred river Ganges, He is known as Ganagadhar. Because of His matted hair, He is dearly called as Jatadhari, by His ascetic followers. As a perfect being He is Siddheshwar. With His Trident in His hands, He became popular as heroic and fearless Trisuladhari. As the world teacher, He was eulogized as Dakhshinamurthi by Adi Shankara in his famous book, "Hymns to Dakhshinamurthi". To the connoisseurs of art, He is well known as Nararaja, master of all dance forms. As the perfect Yogi, He is credited with the source of all knowledge concerning the various yogas. Such was

His prowess and divinity, that Lord Vishnu in His incarnations as Rama and Krishna, worshipped Him with great reverence. He was dearer not only to gods but also to demons.

Many demons like Ravana and Bhasmasur were His ardent followers, to whom He gave several boons out of His unbound love and in the process invited great trouble for Himself and for others. Ever willing to help those who are in distress, He saved the worlds and all the gods by partaking Halahal during the churning of the oceans.

2

Neelkantha

Lord Shiva Drinks Poison

Once, the Devas and the Ashuras had a formidable fight for a very long period. Many Devas and Ashuras died in the fight. The Devas thought that they should prolong their lives by drinking nectar and then continue the war. They approached Brahma with this desire in view. Brahma said: "This work cannot be done by me. It can be done only by Lord Vishnu." Thereupon, Brahma and the Devas approached Lord Hari in Kshirasagara.

Lord Hari asked the Devas and Ashuras to churn the ocean with the help of Mandaragiri as churning-rod and Vasuki, the serpent, as the rope. As they went on churning, poison (Halahala) came out first from the ocean. The terrible poison began to burn the people into ashes. The Devas, the Ashuras and the Rishis began to fly away. Lord Vishnu was not able to destroy the poison. His body also became very black. He ran to Kailas along with the Devas and Brahma to see Lord Shiva. He reported to Lord Shiva all that had happened. Thereupon, Lord Shiva collected the poison and kept it as a drop in the palm of His hand and swallowed it.

Then Lord Vishnu and Brahma requested Lord Shiva to keep it in His neck as a mark of their protection. Lord Shiva did it accordingly. Due to the effect of the poison His throat became blue. From that day onwards Lord Shiva is called by the name 'Neelkantha' (blue-necked one) or 'Kalakantha Murti'. Then Lord Shiva said to them: "If you churn again, you will get nectar and several other things." They all started again the churning of the ocean and obtained nectar and many other things. All the Devas drank the nectar and rejoiced heartily.

Lord Shiva Rides on the Bull

Turning of Chatur-Yuga two thousand times, is a day of Brahma. Such thirty days make a month for him. Such twelve months make a year for him. Such hundred years make the full life of Brahma. The whole life-period of Brahma is a day for Lord Vishnu. After one hundred years, Lord Vishnu also will pass away or merge himself in Parabrahman. All the Andas will perish. On account of the heavy blowing of Prachanda Vayu, the seven oceans will ebb high and cover the whole world. Lord Shiva alone will exist. He will burn everything into ashes through His fiery third eye and then dance.

The Dharmadevata or the Lord of Virtue reflected within himself: "How can I attain immortality? If I approach Lord Shiva only, I can attain this." He assumed the form of a bull, went to Lord Shiva and said: "My venerable Lord! Kindly accept me as Thy vehicle and thus protect me." Lord Shiva agreed to the humble request of the God of Dharma and said: "Conduct the Dharma with four feet in the Krita Yuga, with the three feet in the Treta Yuga, with two feet in the Dvapara Yuga, and with one foot in the Kali Yuga. On account of My grace, you will be endowed with all splendour and powers. You will be always My vehicle. You will be one with Me." When Lord Shiva destroyed Tripuras or the three cities, Lord Vishnu assumed the form of a bull and supported Lord Shiva.

Lord Shiva wears Trident, Deer, Etc.

The Rishis of Darukavana performed a Yajna to destroy Lord Shiva. A cruel tiger came out of the fire. They commanded the tiger to kill Lord Shiva. Lord Shiva killed the tiger and wore the skin around His waist. Then they created a trident, to kill the Lord. Lord Shiva wore it in His hand as His instrument. Then they created a deer with sharp horns, to kill the Lord. Lord Shiva wore it in His left hand. Later on they created countless black cobras to kill the Lord. Lord Shiva wore them as His ornaments. Then they created countless Bhuta Ganas to kill the Lord. Lord Shiva made them as His army. Then they created a Damaru to kill the Lord. Lord Shiva wore it in His hand. Then they created an Asura, Muyalaka, to kill the Lord. Muyalaka marched with the Yajna fire to destroy Lord Shiva. Lord Shiva held the fire as Malu in His hand and kept the Asura under His feet.

Lord Shiva wears the Ganga on His Head

Once upon a time in Mount Kailas, Parvati closed the eyes of Lord Shiva with her hands. Thereupon the sun, moon and fire did not shine forth. This caused terrible havoc in the world. All were enveloped by darkness for a long period. Lord Shiva opened His third eye a bit. The sun, moon and fire began to shine again and all darkness vanished. Parvati was frightened. She removed her hands and dropped down the perspiration from her fingers. This perspiration was turned into ten Gangas with countless branches.

These rivers did much havoc to the world. Thereupon Lord Brahma, Vishnu and Indra ran to Lord Shiva and requested Him to avert this catastrophe. Lord Shiva felt compassionate and brought the whole waters in one hair of His matted locks. Lord Brahma, Vishnu and Indra requested Lord Shiva to give a little of the water of the Ganga for their worlds. Lord Shiva gave them a little. They became Virajanadi in Vaikuntha, Manasa Tirtha in Satyaloka and Devaganga in Indraloka. King Bhagiratha brought down the Ganga from Brahmaloka to save the sixty thousand sons of Sagara.

Lord Shiva's Lila of Begging

The Rishis of Darukavana thought that there was no use in loving and adoring Lord Shiva and they could attain Moksha through the performance of sacrifices. They left off worship of the Lord and did sacrifices vigorously. Then Lord Shiva said to Lord Hari: "Assume the form of Mohini and enter the abode of all Rishis in Daruka forest. They have no regard for Me now. They are treading the wrong path now. We should teach them a severe lesson. Excite their passion and delude them. Destroy their Vratas." He also put on the form of a mendicant-beggar. Thereupon, Lord Hari assumed the form of Mohini and entered the dwelling places of the Rishis in Darukavana. All the Rishis lost their power of understanding and discrimination and followed Mohini under strong excitement of passion.

Lord Shiva entered the Parnakutirs of the wives of the Rishis, sang the Srutis and hymns beautifully and roamed about as a mendicant-beggar. The wives of Rishis became excited and followed Lord Shiva. They entreated Him in a variety of ways to satisfy them. Lord Shiva multiplied Himself and appeared in the mind of each woman. All the wives of the Rishis enjoyed heartily.

They all brought forth in the morning eighteen thousand Rishis with matted locks, Danda and Kamandalu. They all prayed to Lord Shiva. Lord Shiva blessed them to do Tapas in the forest. The Rishis acted accordingly. The Rishis witnessed the condition of their wives and said: "We were deluded by the enchanting Mohini. The mendicant-beggar spoiled the chastity of our wives. Lo! how powerful is lust! Mysterious is Maya!"

Lord Shiva has Uma on His Left

After the Samhara was over, Lord Brahma born of the grace of Lord Shiva thought of creating all beings. He created Shanaka, Sanandana, Sanatkumara and Sanatsujata. They did not enter the life of householders. They developed wisdom and became great Yogis. Lord Brahma went to Vaikuntha and saw Lord Hari and said to Him: "O venerable Lord! I am not able to continue the creation. Shanaka, Sanandana, etc., have become Yogis. They do not wish to become householders. Kindly suggest to me a way to continue my work of creation." Lord Vishnu said: "This is not in my power. Let us go to Lord Shiva who abides in Kailash." Brahma and Vishnu saw the Lord of Kailash and said to Him: "O God of gods! The creative work of Brahma has been stopped as the four Kumaras have become Yogis. Kindly bless him to continue his creative work."

Lord Shiva looked at the left side of His chest. Uma took Her birth from Lord Shiva's left side. Lord Shiva said to them: "Now there is no difficulty in creation. It will continue without any hitch." Thereupon Brahma and Vishnu repaired to their respective abodes. Then Lord Brahma created the worlds through the grace of Lord Shiva. Men and women lived together happily and brought forth offsprings. The whole appearance is Shaktimaya. Lord Shiva witnesses Prakriti's activities.

Lord Shiva wears Elephant's Skin

In days of yore, Gajasura did severe penance. Brahma appeared before him and asked him: "O Gajasura! I am pleased with your penance. What boon do you want?" Gajasura said: "O venerable Lord! Give me prowess and inexhaustible wealth." "I have given you what you desired. But if you fight against Lord Shiva, you will lose your boon." Then Lord Brahma vanished.

Thereupon, Gajasura did Digvijaya, and defeated Devas and Indra. He troubled the Munis and the Rishis also. They repaired to Benares and fell at the feet of Lord Vishvanatha and said: "Gajasura is trying to kill us. Protect us. O Lord! There is no other refuge for us." Gajasura attacked Lord Vishvanatha also. The Lord killed the Asura, tore the skin and wore it as His garment. Then the Devas and the Rishis praised the Lord and became happy.

Ravana and Shiva

Ravana was a great devotee of Lord Shiva. He used to go everyday to Mount Kailas to worship Lord Shiva. He found this very troublesome. He thought within himself to bring the whole mountain to his abode in Lanka in order to save himself from the trouble of a daily journey to Mount Kailas. He began to pull up the mountain which trembled. Parvati, consort of Shiva, got frightened. She embraced Lord Shiva. Shiva pressed Ravana with His toe and sent him down to the nether world.

Hari and Shiva

Hari used to worship Shiva daily with a thousand lotuses. One day one lotus was missing. He plucked out His own eye to make the number of a thousand. Shiva gave Vishnu the Sudarshana Chakra or discus, being very much pleased with His devotion. It is this Sudarshana Chakra which Vishnu always bears. This discus is itself an embodiment of devotion.

Brahma's Boon

A Rakshasa worshipped Brahma and asked Him to grant him the power to destroy the whole world. Brahma was partially unwilling to grant him this boon. He consented in a half-hearted manner and asked him to wait. The Devas, hearing this, were terribly frightened and went to Shiva and told Him everything. Shiva danced in order to delay the granting of this boon by Brahma, to distract him and thus save the world.

Birth of Subrahmanya

The Asura Taraka drove all the gods from heaven. The gods went to Brahma. Brahma said to the gods: "The Asura has acquired his power through my grace. I cannot destroy him. I shall suggest to you one plan. Go to Lord Shiva. He is in Yoga Samadhi. He must be tempted to unite with Parvati. A powerful son will be born unto Him. He will destroy the Asura". Thereupon, Indra asked Cupid (Kama) to go with his wife Rati and his companion Vasanta (the spring), to Mount Kailas, the abode of Lord Shiva. The three persons at once proceeded to Kailas. Spring season appeared there.

All the Rishis were surprised at this. Kama stood behind a tree and shot an arrow at Shiva. At this moment Parvati was worshipping Shiva and offering flowers in His hand. Her hand touched the hands of Shiva. Shiva felt suddenly a thrill of passion and His seed came out. Shiva was wondering what was it that thus distracted Him from His Yoga. He looked around and witnessed Kama behind a tree. He opened His third eye. A fire flashed and burnt Kama to ashes. Shiva's seed was thrown into Agni (God of Fire). Agni was not able to bear it. He threw it into the Ganga. Ganga threw it into a forest of reeds where Subrahmanya, called the reed-born (Sara Janma, Saravana Bhava), was born. Subrahmanya became the commander of the Devas and destroyed the Asura as Brahma intended.

Lord Shiva and Daksha

Daksha goes to attend the sacrifice of the sages who are the progenitors of the world, and not being honoured by Rudra who has come before him, Daksha reproaches Him and leaves the place. Nandisvara in his turn, curses Daksha and other Brahmins. Rudra then leaves the place of sacrifice. Sati, daughter of Daksha, known also by the name Dakshayani, requests Shiva's permission to attend her father's sacrifice, the Brihaspati-sava, and Shiva shows how inadvisable it would be. In spite of her Lord's advice, Sati goes to Daksha's sacrifice. Disregarded by her father and enraged at finding no offering made to Rudra, she praises the greatness of her Lord, censures Daksha and by the Yoga method casts off her body.

Hearing of Sati's casting off her body, Shankara creates in His anger Virabhadra out of His Jata (or matted locks) and causes Daksha's death. Informed by the gods of the destruction of Daksha, Brahma pacifies Rudra, and Daksha and others are revived. Being pleased with Brahma's praise, Shiva goes along with the gods to the scene of sacrifice; Daksha and others are revived; Vishnu issuing forth from the sacrificial fire is praised by Daksha and others. After the closing ceremony of ablutions, the gods return home. Maitreya describes the benefit of listening to this story, the birth of Sati as Parvati, and that of Skanda.

Dakshinamurti

On the Mount Kailas, with Parvati Devi by His side, Lord Shiva was sitting in a hall beautifully decorated by precious stones. At that time, Devi worshipped the Lord and requested Him to alter the name of Dakshayani given to her before, for being the daughter of Daksha. This Daksha was killed by Lord Shiva for his disrespect and arrogance. On hearing this request Lord Shiva ordained that Devi should be born as the daughter of Parvata Raj who was doing rigorous Tapas for getting a child. He also told Parvati that He would come over to her and marry her.

Thus ordained, Parvati Devi was born as the child of Parvata Raj and since her fifth year, began to do rigorous Tapas for being the bride of Lord Shiva. During the absence of Devi, when Lord Shiva was alone, the sons of Brahma, who are sages Shanaka, Sanandana, Sanatana and Sanatkumara, came to have Darsana of Lord Shiva and prostrated before Him. They entreated the Lord to teach them the way to remove Avidya and attain salvation. They expressed that in spite of the vast study of scriptures they had no internal peace and they were in need of learning the inner secrets, by knowing which they could attain salvation.

Lord Shiva, hearing this appeal made by the sages, assumed the form of Dakshinamurti and remaining as the Guru Supreme, began to teach them the inner secrets by keeping Mouna and showing the Chinmudra by His hand. The sages began to meditate on the lines shown by the Lord and attained the state of inexpressible and illimitable joy. Thus Lord Shiva came to be known as Dakshinamurti. May the blessings of Lord Dakshinamurti be upon us all! May you all dive deep and enjoy the everlasting Peace and Bliss through His Grace!

Tripura Samhara

This occurs in the Karna Parva of the Mahabharata. In times of yore, there was a war between the Devas and the Ashuras. The Ashuras were defeated in the battle. The three sons of the Asura, Taraka, wanted to take revenge on the Devas. They performed rigorous penance and obtained a boon from Brahma, the creator. By virtue of this boon, they could not be conquered by any one save somebody who could destroy their three castles by a single arrow. They made three castles, one of gold in heaven, the second silver in air, and the third of iron on earth.

They began to oppress the gods and the Rishis. Thereupon, all the gods made a complaint to Brahma. Brahma replied that no one save Mahadeva who knows Yoga and Sankhya, through particular penance, could vanquish these Ashuras. All the gods approached Mahadeva and provided Him with a chariot out of all forms of the universe. They also supplied Him with a bow and arrow, the constituent parts of which were Vishnu, Soma and Agni. They requested Mahadeva to discharge the arrow against the three castles.

Brahma became the charioteer. Mahadeva discharged the arrow against the three castles. The three castles fell down in the twinkling of an eye. Then all the gods eulogised Mahadeva and departed to heaven. Mahadeva said to the gods that He Himself could not destroy the Ashuras, as they were very strong, but that they themselves would be able to vanquish with the help of half His strength. The gods replied that they could not bear half His strength and that He should take up the task with the help of half their strength. Mahadeva agreed to this. He became stronger than all the gods. Hence He was called Mahadeva or the great God.

Lord Shiva Curses and Pardons Nakirar

Once, a Pandya king of Madurai felt that his queen's hair had some kind of natural fragrance. A doubt arose in his mind as to whether human hair could have natural fragrance, or could be rendered fragrant only through association with flowers or scents. He went to the Sangham or the Tamil Academy the next day, suspended a bag containing one thousand gold pieces and said to the poets that anyone who would write a poem clearing the doubt he entertained in his mind would get the gold pieces as a prize.

Many poets composed poems but they were not able to satisfy the king. Darumi, a Brahmin priest in the temple, was extremely poor. He requested Lord Shiva thus: "O all-merciful Lord! I am very poor. I wish to marry now. Relieve me of my poverty. Help me to get these gold pieces now. I take refuge in Thee alone". Lord Shiva gave him a poem and said: "Take this poem to the Sangham. You will get the gold pieces". The king was immensely pleased with the song as it cleared his doubt, but the Sangham poets did not accept it. Nakirar, one of them, said that there was a flaw in the poem.

The poor priest was greatly afflicted at heart. He came back to the temple, stood in front of the Lord and said: "O Lord! Why did You give me the poem which contained a flaw? Nobody will take You for the Lord. I feel very much for this". The meaning of the poem is: "O fair-winged bee! You spend your time in gathering flower-dust. Do not speak out of love, but speak out of truth. Is there any among the flowers known to you that is more fragrant than the hair of this damsel who is most loving, is of the colour of the peacock and has beautiful rows of teeth!"

Thereupon, Lord Shiva assumed the form of a poet, went to the Sangham and asked: "Which poet found out flaw in the poem?" Nakirar said: "It is I who said that there is a flaw." Lord Shiva asked: "What is the flaw?" Nakirar said: "There is no flaw in the composition of words. There is flaw in the meaning." Lord Shiva said: "May I know what defect is there in the meaning?" Nakirar said: "The hair of a damsel has no natural fragrance.

It gets the fragrance from association with the flowers." Lord Shiva said: "Does the hair of Padmini also possess fragrance by association with the flowers?" Nakirar said, "Yes." Lord Shiva said: "Does the hair of celestial damsels also possess fragrance by association with the flowers?" Nakirar replied, "Yes. Their hairs become fragrant by association with Mandara flowers." Lord Shiva said: "Does the hair of Uma Devi who is on the left side of Lord Shiva whom you worship possess fragrance by association with the flowers?" Nakirar replied: "Yes. Quite so." Lord Shiva slightly opened His third eye. Nakirar said: "I am not afraid of this third eye. Even if you are Lord Shiva, even if you show eyes throughout your whole body, there is flaw in this poem."

The fire from the third eye of Lord Shiva fell upon Nakirar. Nakirar was not able to bear the heat. At once he jumped into the neighbouring lotus-tank to cool himself. Then all the poets approached Lord Shiva and said: "O Lord! Pardon Nakirar." Lord Shiva appeared before Nakirar. Through the grace of Lord Shiva his body was rendered cool. He repented for his mistake and said: "I pointed defect even for the hairs of Uma Devi. No one but the Lord can pardon me." He sang a song with intense devotion. Lord Shiva entered the tank and brought him to the shore. Then Nakirar and other poets gave the purse of gold to Darumi.

Know Your Guru

Devotion to the Lord dawns in the heart of a man who has done virtuous actions in his previous births without expectation of fruits and egoism or the idea of agency. Devotion leads to knowledge of the Self (Jnana) and through Jnana, he attains

Moksha or the final emancipation. In days of yore, Virasindhu was the king of Kalinga country. He did severe Tapas, meditation and Yoga in his previous birth, but did not attain Moksha. He became a Yoga-Bhrashta as he had to enjoy the fruits of some residual Karmas in this last birth of a King. He took his birth as the son of a king. When he attained the proper age, he was crowned as the king of Kalinga. He ruled the kingdom for a period of ten years.

Owing to the force of previous spiritual Samskaras, and the grace of the Lord, discrimination and dispassion dawned in his heart. He reflected within himself: "I am doing the same acts of eating, drinking and sleeping. My several forefathers who ruled the country, had been reduced to dust. I have no peace of mind despite my wealth and dominion. I should get a Guru and obtain initiation from him to get knowledge of the Atman and reach the abode of immortality and eternal bliss.

King Virasindhu sent invitation to all Pandits, Sannyasins, Sadhus and Mahatmas. He wrote in the letter thus: "I will give half of my dominion to that supreme Guru who will give me the right initiation and make me realise the Self. If he fails to do so he will be put in the prison."

Many Pandits and Sadhus saw the king. One gave him Taraka Mantra, another gave Panchakshara, a third gave him Ashtakshara, but no one was able to satisfy the king. He put them all in the prison. He had initiation into these Mantras in his previous births.

King Virasindhu became very restless as he did not obtain his Guru. Lord Shiva assumed the form of an ordinary coolie. He had a very black complexion. He wore some rags and appeared before the king. The king went in advance to receive him. Through the grace of the Lord, he came to know that this coolie was none other than the Lord.

The coolie raised his hand towards the king and said "stop", and vanished immediately. The king understood that the Guru has commanded him to stop the mind and control its movements. The king closed his eyes, while he was standing, and did not allow the mind to think of the sensual objects. He controlled all the modifications of the mind. It was easy for him as he had practised Yoga and meditation in his previous birth. He entered into Nirvikalpa Samadhi and became like a statue. He did not open his eyes.

The ministers were waiting in the Durbar Hall for hours together. King Virasindhu did not open his eyes for days together. Then the ministers reflected: The king is in Samadhi now. We do not know when he will return from the Samadhi. We will have to manage the affairs of the State." They removed the ring from his finger and used it for putting the seal in the papers of the State. The king opened his eyes after six years, and asked the ministers: "Where is my Guru?" The ministers replied: "O venerable king! The Guru said a word to your majesty and vanished at once. You are standing here like a statue for the last six years. We are conducting the affairs of the state with the aid of your ring. Here is the seal of your ring in all the registers and papers."

The king was struck with awe and wonder. He thought within himself: "Six years have passed like a second. I enjoyed supreme bliss. I have no desire to rule the State after tasting the supreme bliss." He left the palace and entered the forest and sat in Samadhi.

The force of Yogic Samskaras of the king that was generated by the practice of Yoga in his previous birth, helped him in the attainment of the beatitude in this birth. Those who have not much piety and religious inclination in this birth should do Japa, Kirtan, meditation, and study of religious books. They should live in the company of sages. They will develop good religious Samskaras. This will be a valuable asset for them in the next birth. They will start the practice of Yoga in the next birth in the early age.

It is difficult to say in what form the Guru or the Lord will appear before you to initiate you. He may come in the form of a leper as Sri Hanuman did, or in the form of an untouchable as Lord Krishna did, or in the form of a groom as Lord Shiva did. The aspirants must be very careful and vigilant to detect the Lord in the various forms which He assumes.

The Twenty-five Lilas of Lord Shiva

The following are the 25 Lilas (sportive plays) or manifestations of Lord Shiva:

> *1. Wearing of moon on the head, 2. Living with Uma Devi, 3. Riding on ox, 4. Tandava dance with Kali, 5. Marriage with Parvati, 6. Begging, 7. Burning of Manmatha or the God of Love, 8. Victory over Yama or the God of Death, 9. Burning of Tripuras, 10. Killing of Jalandarasura, 11. Killing of Gajasura, 12. Incarnation of Virabhadra, 13. Harihara, 14. Ardhanarishvara, 15. Transforming into Kirata, (hunter), 16. Assuming the form of Kankala, 17. Blessing Chandisvara, 18. Drinking poison, 19. Giving of Chakra to Lord Vishnu, 20. Destroying of obstacles, 21. Having sons of Uma Devi with Him, 22. Becoming Ekapada Rudra, 23. Being in easy pose (Sukhasana), 24. Assuming the form of Dakshinamurti, and 25. Assuming Linga form.*

Rudra Upanishad

Now, at this moment, I take refuge in that Pure State of the Supreme Absolute which can be known by the Vidya, called the Rudra Hridaya Upanishad.

After prostrating before the celebrated form of Sri Mahadeva-Rudra in his heart, adoring the sacred Bhasma and Rudraksha and mentally reciting the great Mahavakya-Mantra, Tarasara, Sri Suka asked his father Sri Vyasa Maharishi, thus:

> *Who is the real God of gods? In whom are all these existences established? By worshipping whom, can I please the Devas in whole?*

Hearing these words, Sri Veda Vyasa replied thus:

> *Rudra is the embodiment of all Devas. All devas are merely different manifestations of Sri Rudra Himself. On the right side of Rudra, there is the sun, then the four-headed Brahma, and then three Agnis (fires). On the left side, there exist Sri Umadevi, and also Vishnu and Soma (moon).*

Uma Herself is the form of Vishnu. Vishnu Himself is the form of the moon. Therefore, those who worship Lord Vishnu, worship Shiva Himself. And those who worship Shiva, worship Lord Vishnu in reality. Those who envy and hate Sri Rudra, are actually hating Sri Vishnu. Those who decry Lord Shiva, decry Vishnu Himself.

Rudra is the generator of the seed. Vishnu is the embryo of the seed. Shiva Himself is Brahma and Brahma Himself is Agni. Rudra is full of Brahma and Vishnu. The whole world is full of Agni and Soma. The masculine gender is Lord Shiva. The feminine gender is Sri Bhavani Devi. All the mobile and immobile creation of this universe, is filled up with Uma and Rudra. The Vyakta is Sri Uma, and the Avyakta is Lord Shiva. The combination of Uma and Shankara is Vishnu.

Hence everybody should prostrate to Sri Maha Vishnu with great devotion. He is the Atman. He is the Paramatman. He is the Antaratman. Brahma is the Antaratman. Shiva is the Paramatman. Vishnu is the Eternal Atman of all this universe. This whole creation of Svarga, Martya and Patala Lokas is a big tree. Vishnu is the top portion (branches) of this tree. Brahma is the stem. The root is Lord Shiva.

The effect is Vishnu. The action is Brahma. The cause is Shiva. For the benefit of the worlds. Rudra has taken these three forms.

Rudra is Dharma. Vishnu is the world. Brahma is Knowledge. Therefore, do Kirtan of His name, 'Rudra', 'Rudra'. By singing like this, the hallowed name of this great Lord, all your sins will be destroyed.

Rudra is man. Uma is woman. Prostrations to Him and Her.

Rudra is Brahma. Uma is Saraswati. Prostrations to Him and Her.

Rudra is Vishnu. Uma is Lakshmi. Prostrations to Him and Her.

Rudra is Sun. Uma is shadow. Prostrations to Him and Her.

Rudra is moon. Uma is star. Prostrations to Him and Her.

Rudra is day. Uma is night. Prostrations to Him and Her.

Rudra is Yajna. Uma is Vedi. Prostrations to Him and Her.

Rudra is Agni, Uma is Svaha. Prostrations to Him and Her.

Rudra is Veda. Uma is Shastra. Prostrations to Him and Her.

Rudra is tree. Uma is creeper. Prostrations to Him and Her.

Rudra is scent. Uma is flower. Prostrations to Him and Her.

Rudra is meaning. Uma is word. Prostrations to Him and Her.

Rudra is Linga. Uma is Pitha. Prostrations to Him and Her.

The devotee should worship Sri Rudra and Uma with these Mantras referred to above. O my son, Suka! With these hymns, you should meditate on the Eternal Para-Brahman, which is beyond the reach of the senses, which is pure Existence, knowledge and Bliss and which cannot be understood either by the speech or by the mind. After knowing this, there is nothing more to be known, because everything is the form of That, and there is nothing separate from That.

There are two Vidyas to be known. They are Para and Apara. Apara Vidya is the embodiment of the four Vedas and their six Angas. They do not deal with the Nature of the Atman. But the Para Vidya is called the Moksha-Shastra. It deals with that supreme philosophy of the Absolute Truth, ununderstandable, impersonal, Nirguna, Nirakara, without ears, without eyes, without hands, without feet, eternal, omnipresent, imperishable, and knowable by the intelligent daring sages.

From that Lord Shiva who performs a terrible penance in the form of Supreme Jnana-Marga, this whole world is created which is the food of the mortals. This world is Maya. It seems to appear just like a dream. It is superimposed on the Lord just like a rope on a serpent. This is the eternal Truth. There is no creation in reality. All is absolute. All is Truth. Knowing this, one is liberated at once.

Only through Jnana, you can get rid of this Samsara. Only through Jnana, you can understand this existence and never through Karma. Understand this through the guidance of a Brahmanishtha-Srotriya Guru. The Guru will give the disciple all the necessary knowledge of Brahman, the Absolute. By cutting off the bondage of Ajnana or Avidya, one should take refuge in Lord Sadashiva. This is the real wisdom to be understood by an aspirant seeking after Truth.

The Pranava is the bow. The Atman is the arrow. The Para-Brahman is the target. Just like the arrow, the Atman will become one with Brahman.

But all these three, the bow, the arrow and the target are not different from that Sadashiva. There do not shine the bodies of the sun, moon or the stars. There does not blow the wind, there do not exist many Devatas. He, the One Lord only exists. He only, the Purity of purities, shines for ever and ever.

There are two birds in this body, the Jiva and the Paramatman. The Jiva eats the fruit of his Karmas, but the Paramatman is untouched by anything. The Paramatman is only the Sakshi. He does not do anything. He only assumes the form of the Jiva through His Maya, just as the Akasa inside a pot seems to be different from the Akasa outside and assumes the form of the pot. In reality all is Shiva, Advaita, the One Absolute. There is no difference of whatever kind.

When all is understood to be One, Omkara, the Absolute, there is no sorrow, there is no Maya. Then the attainment of the Advaita-Paramananda is very easy. Think that you are the basis of all this universe, you are the One, Kevala, Sat-Chit-Ghana. All people cannot understand this Truth. Those devoid of Maya can know this secret. After knowing this, the Atman does not move towards any place at any time. It becomes one with the Absolute, just like Ghatakasa with Paramakasa. Just as Akasa does not move anywhere, similarly this Atman does not have any movement. It becomes one with OM.

One who knows this great secret Truth is the real Muni. He becomes the Para-Brahman Itself. He becomes Sachidananda. He attains permanent peace.

Dust of Gurudev's Feet

Venkatesananda

Sundaramurthi Nayanar: Sundaramurthi Nayanar flourished in the 8th century. He was a great devotee of Lord

Shiva. He is one of the Tamil Samaya Acharyas (four Tamil religious Teachers).

Sundaramurthi Nayanar sang the glories of Lord Shiva at all the sacred places that he visited. These hymns are called Thevaram. They have been collected into a book-form. All devotees sing the Thevaram even today. The hymns sung by Sundarar, Appar or Tirunavakkarasu, and Tirujnana Sambandar are called Thevaram. The hymns of Manickavachagar are called Thiruvachagam.

Sundarar had the Sakhya Bhava or the attitude of a friend towards the Lord. He freely demanded of the Lord whatever he wanted. He did not do so with selfish desire, however. Whatever he asked for was for the sake of those who were dependent on him. He lived only eighteen years.

Sundaramurthi Nayanar was born in Thiru Navalur where the entire atmosphere was full of spiritual vibrations and Shaivism was well established. In this place, there lived a pious, devout and respected Brahmin by name Sadaiyanar whose ancestors were all ardent devotees of Lord Shiva. Isaignaniar was his dutiful wife. She gave birth to a divine child whom the parents named 'Nambi Arurar' after its grandfather.

In his previous incarnation Arurar was Alala Sundarar, who was an ardent devotee of Lord Shiva. When the Milky Ocean was being churned by Devas and Ashuras, a deadly poison began to spread on the surface of the ocean threatening the existence of all beings. Then Alala Sundarar collected that poison in his hand and gave it to Lord Shiva Who drank it for the protection of the world. Hence, Sundarar got the word Alala (for Halahala, the poison) prefixed to his name.

Once when Alala Sundarar was living by the side of Lord Shiva in the Mount Kailas, serving the Lord and bringing flowers from the garden for His worship, he cast a lustful look at Aninditi and Kamalini, the attendants of Goddess Parvathi who had also gone to the garden to collect flowers for the divine Mother's worship. They, too, fell in love with him. Lord Shiva, through His divine vision, understood all that had happened in the garden. He called Alala Sundarar and said: 'Sundarar, since you fell in love with these girls, you and they, too, will go down to the earth and take a human birth. You will marry them and enjoy the pleasures of the world.' Sundarar wept bitterly, regretting his folly which had resulted in his separation from the Lord. He prayed to the Lord: 'Oh Lord! It is due to my evil thought that I have to undergo this separation from Thee. I am afraid lest I should be steeped in ignorance and forget Thee. Oh Lord of mercy! Let this not happen to me. Oh Lord of compassion! Dispel my ignorance soon and take me back to Your lotus feet.' Lord Shiva granted this wish.

There was another cause for Sundarar's human birth. To an ordinary man it may appear that Sundarar was a victim to lust, even in the divine realm of Kailash. It was not so. Sundarar was only an instrument in the hands of God. It was Lord Shiva's wish that Sundarar should sing Tiru Thonda Thogai for the benefit of mankind. So, Lord Shiva entered his mind and created a desire for these two girls.

Also, the Lord wanted to teach mankind a great lesson. Lust is extremely powerful. It can delude even a great devotee of the Lord like Sundarar, if he is not ever vigilant. Maya's charms are powerful. Unless this evil quality is burnt, the Jiva cannot reach Shiva. Yet another lesson. The lustful eye was the cause of Sundarar's downfall. But, when it is used in the service of the Lord (for looking at the holy shrines, holy images of God, saints, and study of scriptures) the very same organ will help towards our emancipation.

Sundarar was, therefore, born as Arurar. The king of that place, Narasinga Munaiyar, happened to see the beautiful child. He liked him. He wanted to bring him up himself and asked for the parents' permission, Sadaiyanar, whose mind was full of dispassion and who was not attached to anything in this world, immediately complied with the king's wish. As we shall see later, he and his devout wife are also regarded as Nayanars.

The boy grew up under royal care. At the proper age, the parents wanted to get their son married. Sadaiyanar sought Sandakavi Shivachariar's consent to obtain his daughter's hand for his son, Arurar. Shivachariar gladly agreed. But, the wedding was not to take place.

Just when the ceremony was to begin, an old Brahmin, with sacred ashes on his body, Rudraksha around his neck and matted locks on his head appeared and said: 'This man, Arurar, is my bond-slave. I have a document to that effect executed by his grandfather. He cannot marry.' This put an end to the ceremony. Sundarar and the Brahmin left the place. The young bride fixed her mind on the holy feet of Sundarar, shed her mortal coil and attained the immortal abode of Lord Shiva.

Sundarar and the old man had a heated argument. Sundarar asked him: 'Who are you and from where have you come?' To which the Brahmin replied: 'I belong to Tiruvennai Nellur.' Sundarar called him a liar and said: 'Come, let us go to Tiruvennai Nellur and get this dispute settled by the wise men there.'

At Tiruvennai Nellur, before an assembly of wise men the old Brahmin produced the document which read as follows:

> *I, Aruran, the Adi Saivite of Tirunavalur, execute this bond of slavery with heart and soul. I and my progeny for all time to come are bond-slaves to Pithan of Tiruvennai Nellur, and we are bound to serve him by all means.*

Pithan means Lord Shiva who delights to be called a 'mad man', to exemplify the state of the highest Yogi whose behaviour resembles that of a mad man but who teaches us that there is nothing in this world worth taking any notice of and the worldly 'wise men' are all mad people in truth.

After examining the witnesses cited in the document and verifying the grandfather's signature, the assembly confirmed the old man's claim. Sundarar had to accept it as God's will. Followed by all of them the Brahmin entered the temple of Tiru Arul Turai on the pretext of showing them his house, and promptly vanished. Arurar understood that it was the Lord Himself who had appeared as the old man to save him from the shackles of Samsara. He was afflicted very much at heart that he had not recognised Him earlier. He cried aloud. The

Lord appeared before him and blessed him: 'Oh noble soul. You are already My Bhakta. You were in My Abode in Kailash before this birth as a man. A wrong thought made you take this birth. Now I have Myself come to save you.'

Because Sundarar had quarrelled with Him, the Lord Himself called him Vanthondan (the devotee who used harsh words) and asked him to sing His glories. 'My clear child, you called Me Pithan (madman) during your quarrel. So, begin with this word and compose a poem.' Sundarar did so: the result was that inspiring poem Pitha Pirai Soodi. The Lord Himself came to be known as Taduthatkonda Iswar (the Lord prevented and saved him from Samsara).

Sundarar later visited a number of holy places and sang the praise of the Lord in all of them. He came to Adigai Virattanam, the sacred place where Appar served the Lord Viratteswarar and was blessed. Sundarar did not like to place his foot on the sacred ground and so stayed on the outskirts of the village.

That night when Sundarar was asleep, the Lord in the guise of an old man entered the Mutt. He lay down close to where Sundarar was sleeping and pretended to sleep. He then placed His feet on the head of Sundarar. When Sundarar objected to this, the old man apologised. Sundarar went over to another corner of the room. There, too, the old man repeated the same action. Sundarar did not lose his temper. He calmly asked him for his identity and explanation for the abnormal behaviour. 'Oh friend, don't you know me?' asked the Lord and disappeared. Sundarar realised that it was again the Lord Himself. Since he had not gone into the place for His Darshan, the Lord Himself had come out to where the devotee was! Sundarar prayed: 'Oh Lord! How kind and merciful You are! Even devotees who are well versed in Vedas and Agamas cannot touch Your feet. Out of love towards this poor creature, You left Your abode and came here to bless me with Your Holy Feet.'

Sundarar again continued his pilgrimage. At Tillai (Chidambaram), he went into a trance even as he saw the temple tower. In his ecstasy he rolled on the ground and shed profuse tears of love. He had the Darshan of Lord Nataraja. A heavenly voice commanded him to go to Tiruvarur. He then visited many other shrines and came to Tiruvarur. The Lord appeared to the Brahmins of Tiruvarur and asked them to receive Sundarar with due honours. They did so. As Sundarar was worshipping the Lord in the temple, he heard a heavenly voice: 'Sundarar! I have made you My friend. I prevented you from getting married. Hereafter you will appear for ever as a bridegroom and sport on earth.' Immediately, Sundarar became a handsome bridegroom. People called him Tambiran Thozhar (friend of God).

In Tiruvarur, there was a chaste woman by name Paravayar who was none other than Kamalini, the attendant of Parvathi in Kailash. Daily she would go to the temple and worship the Lord with faith and devotion and sing His glories. One day, she came to the temple, as usual, with her friends, to worship the Lord. At the same time, Sundarar, with his devotees entered the temple. Prompted by past Karma, Sundarar was attracted by Paravayar's beauty. He wanted to marry her, and entered the shrine of the Lord with this thought. The Lord was his friend, and so, he expressed his desire to Him!

Paravayar who had seen Sundarar in the temple also fell in love with him and wanted to marry him.

The marriage was pre-ordained by Lord Shiva Himself and it was now His duty to bring it about. He appeared to both of them in their dreams and told them that they would get married. He also commanded His devotees in dream to arrange for the wedding of Paravayar and Sundarar the very next day. This was done accordingly, to the joy of both Paravayar and Sundarar.

One day, Arurar went into the temple and found a number of devotees of the Lord there. He wanted to sing their glories. The Lord Himself sang the first line of the famous poem Tiruthonda Thogai and by His grace, Sundarar completed it.

During his stay at Tiruvarur, a Vellala by name Kundaiyur Kizhar who was very highly devoted to him, was regularly supplying Paravayar with enough grains and groceries for the maintenance of Sundarar and the devotees. Suddenly there was famine in the district and people suffered for want of food. Kundaiyur Kizhar was also affected. He was afflicted at heart because he could not supply the needs of Sundarar. Lord Shiva appeared in his dream and promised enough grain! Kubera, the God of wealth, did the needful, as commanded by the Lord. The next morning, Kundaiyur Kizhar found huge heaps of grain. At the same time the Lord appeared before Sundarar and informed him of the incident.

At once Sundarar left for Kundaiyur to meet Kizhar. They met half-way. Sundarar saw the heaps of grain at Kundaiyur and knew that it was His Lila. He went to a nearby Koili and sang the praise of the Lord, and entreated Him to have the grain removed to Tiruvarur. A celestial voice immediately assured him of this. Sundarar returned to Tiruvarur and informed Paravayar of all that happened. That night the Bhuta Ganas, the servants of Lord Shiva, removed the heaps of grain and filled the entire town of Tiruvarur with it! Paravayar offered repeated prostrations to the Lord and sang His glory. She asked the people to take the grain to appease their hunger. Thus the famine came to an end. All the people glorified the Lord and Paravayar.

Kotpuli Nayanar of Tirunattiyattankudi, the Commander-in-chief of a Chola King, and an ardent devotee of Lord Shiva came to Sundarar and entreated him to grace his house with his presence. Sundarar agreed to this and went. After worshipping Sundarar, Kotpuli Nayanar prostrated himself at Sundarar's feet along with his two daughters, Singadiyar and Vanappahaiyar, and pleaded that Sundarar should marry the two daughters. Sundarar, however, placed them on his lap and fondled them, treating them as his own daughters. Then Sundarar went to the temple and sang in praise of the Lord, a song in which he called himself Singadiappan, since he took Singadiyar as his daughter.

Sundarar then returned to Tiruvarur. It was Paravayar's custom to distribute plenty of money and other articles in charity on Panguni Uttaram, a festival day. Sundarar went to Tirupugalur and prayed to the Lord to give him gold for the sake of Paravayar. That night he slept there with a few bricks as his pillow. The next morning, he woke up to find that all the bricks had been converted into gold. Sundarar was surprised at this miracle of Lord Shiva and sang His glory and returned to Tiruvarur. On the way he had a vision of the Lord at Tiru Panaiyur.

After visiting many holy places again and singing hymns in praise of the Lord, Sundarar came to Tiru Pachilasramam. There he worshipped the Lord and asked for a gold coin. He did not get it immediately. He sang a Padigam (song) and the Lord at once gave him a heap of gold. The Lord was so fond of hearing Sundarar sing.

Then, Sundarar left for Vridhachalam, visiting a number of holy places on the way. He had omitted Tiru Koodalaiyarrur. So, the Lord came to him as a Brahmin of whom Sundarar enquired the way to Vridhachalam. The Brahmin led the way up to a certain distance, and then suddenly disappeared. It was close to Tiru Koodalaiyarrur which Sundarar now visited and sang a song in praise of the Lord there.

Then Sundarar came to Vridhachalam. He worshipped the Lord, and sang a Padigam expressing his desire for gold coins. The Lord gave him 12,000 pieces of gold. Sundarar prayed to the Lord to remove these gold pieces to Tiruvarur. The Lord asked him to throw them into the river Manimukta and to receive them back at Tiruvarur. Sundarar did so, keeping a piece for identification. On return to Tiruvarur, Sundarar and Paravayar went to the tank to get back the gold pieces. Sundarar dived into the eastern side of the tank and searched for the gold, as though he had put them there. He could not find them. Sorely afflicted at heart, he sang a song. That was what the Lord wanted. Sundarar got the gold. All were amazed. But, on identification, it was found that the gold was inferior in value to the piece that Sundarar had kept back with him. He sang a song: and the Lord restored to them their original value. So fond was He of hearing Sundarar sing a song.

Sundarar went out on another pilgrimage again. On the way, he was afflicted with hunger and thirst. The Lord Who is the Indweller of our hearts, erected a water-shed and was waiting for Sundarar there in the guise of a Brahmin. Sundarar and the devotees entered the shed, singing the Panchakshara. The Brahmin offered him food and water and asked him to rest awhile. All of them appeased their hunger, but the quantity of the food remained the same. When they were resting, after food, the Lord disappeared. They knew that it was none other than the Lord Himself. Sundarar sang a song alluding to this incident.

On another occasion, soon after this, while on a visit to Tirukachur, Sundarar went to the temple, worshipped the Lord and was resting outside the temple, feeling hungry. The Lord understood it: and so, in the guise of a Brahmin came to Sundarar and said: 'It appears that you are hungry. Please wait here. I will give you food.' The Lord at once went out in the scorching sun, begged from each and every house, and offered the food so obtained, to Sundarar. As Sundarar and the devotees were eating, the Brahmin disappeared: and they understood that it was the Lord Himself. Sundarar sang a song alluding to this incident, revealing the Lord's supreme mercy.

Later on, he went to Tiruvotriyur and stayed there for some time, worshipping the Lord there. Aninditiyar, the other maid-servant of Parvathi in Kailash, who had also taken a human birth, was now Sangilyar in Jnayiru in Thondai Nadu. Her father was Jnayiru Kizhar, a Vellala by caste. He was also a staunch devotee of Lord Shiva. Sangiliyar was devoted to Parvati from her very childhood. Once her parents mentioned that she should get married, but the very word 'marriage' made her faint. Later, a respectable Vellala wanted to marry her. He sent some people to approach the girl's father. Jnayiru Kizhar did not like even to speak to his daughter about it. He sent them away with an evasive reply. Soon after, the boy who wanted to marry Sangiliyar, and the party that went to negotiate,—all of them died. When Jnayiru Kizhar heard this, he understood the greatness of his daughter. He took her to Tiruvotriyur and built a small Ashram for her there.

It was part of Sangiliyar's Sadhana to make garlands for the Lord in the temple. She regularly visited the temple and worshipped the Lord. One day Sundarar and the devotees went to the temple. After the worship, they came to the place where some devotees were making garlands for the Lord. Sangiliyar was also there. Sundarar was attracted by her beauty, due to past Samskaras. He wanted to marry her, and expressed this wish to the Lord. The Lord promised to fulfil his wish.

The Lord appeared in Sangiliyar's dream and said: 'Oh noble soul, I am highly pleased with your devotion. Now I tell this for your own good. Sundarar wants to marry you. He is My friend. He asked Me to arrange the marriage. So, marry him. You will be happy.' Sangiliyar prostrated before the Lord and said: 'Oh Lord, I will obey Your command and marry him. But, he may desert me since he is already married.' The Lord asked her to get a promise from Sundarar that he would not part from her under any circumstance.

Then the Lord appeared before Sundarar and said that Sangiliyar had agreed to marry him on condition that he would not part from her. Sundarar said: 'Oh Lord, how can I agree to this condition since I am constantly moving about visiting many holy shrines? But, if You so desire, then assure me that You will withdraw Your presence from the Lingam in the temple and will take Your abode in the nearby tree, when I take the oath before Your image.' The Lord granted him this wish and disappeared. He again appeared before Sangiliyar and said: 'Oh noble soul, Sundarar has agreed to your condition. But, ask him to make this promise, not before the Lingam in the temple, but in front of the nearby tree.'

The next morning Sangiliyar came to the temple. Sundarar was waiting there for her. Sangiliyar's friends told him that she wished the promise to be given in front of the nearby tree. He was taken aback, but accepted the proposal. The promise was given. The marriage was immediately solemnised.

In Tiruvarur, Vasanta Utsavam was being celebrated on a grand scale. Sundarar remembered the festival and longed to go there. He also recollected that Paravayar would sing and dance there in front of the Lord. At the same time, he could not part from Sangiliyar. For a long time, he struggled between the two conflicting duties. Finally, he decided to leave.

When he crossed the border of Tiruvotriyur, he suddenly lost his eye-sight and fell down on the ground. The Lord is impartial. None can escape the operation of the Law of Divine Justice. Sundarar slowly regained his consciousness. Immediately he realised his fault and prayed to the Lord for forgiveness, asking for the grant of the eye-sight. 'Oh Lord, I take complete refuge in You. I always repeat Your Name. Even when I fell down losing my eye-sight, I remembered You only. Oh Lord of Mercy, even if I commit a crime, is it not Your duty to forgive me? Oh Lord of Compassion, save me.'

In spite of the blindness, however, his thirst for the Darshan of the Lord at Tiruvarur did not abate. With the help of some people on the way he came to Tiru Mullaivayil. Here, again, he prayed to the Lord to give him eye-sight. At Tiruvembakkam he again prayed to the Lord in a similar strain. In the temple, he prayed and asked: 'Oh Lord Who accepted me into His fold, prostrations unto Thee. Oh Lord Who cleverly played a trick on me, are You inside the temple?' The Lord gave a stern reply: 'I am here; you can go.' and gave him a blind-man's stick. This attitude of indifference on the part of the Lord pained Sundarar and he pleaded for mercy. 'Oh Lord of Mercy, have I not taken You as my sole refuge and support? I committed a mistake thinking that You will pardon me.

You are even indifferent to public criticism. Will they not accuse You for turning a deaf ear to a devotee who is sincerely weeping at Your feet, accepting his fault and craving for pardon? Oh Lord, can You not understand suffering? Like a loving child that has been separated from its mother for a long time and wants to hug her, I have come to You: but, instead, You treat me like a stranger. Oh Lord, You deceived me, who asked You to remain for a while near the tree, by cleverly informing Sangiliyar of the same and asking her to get the promise from me near the tree. Oh Lord, You gave me Sangiliyar and all the pleasures. But, now You give me the blind-man's staff and say 'You can go'. Oh Lord, am I unfit to receive Your mercy? Pardon me and relieve me of my sufferings.'

Then, completely resigning himself to God, Sundarar came to Conjeevaram, after visiting many holy places on the way. He worshipped Mother Kamakshi and expressed his sufferings to her and pleaded to her to relieve him of his sufferings.

Sundarar then worshipped Lord Ekambareswarar. Mother Who is seated on His left side, had already been moved by Sundarar's prayers and wanted to shower Her grace on him. Lord Shiva understood this. He at once restored vision to Sundarar's left eye. In ecstasy he rolled on the ground, shedding profuse tears of love.

After spending some days there, singing the glories of the Lord, Sundarar proceeded to go to Tiruvarur. At Tiruvavaduthurai, he again prayed to the Lord to forgive him and restore sight to the other eye, too. Sundarar then came to Thiruthurithi. The Lord asked him to take a dip in the northern tank there. Sundarar did so, and came out of it. To the surprise of all, his body had become as bright as polished gold. People were amazed at this change. Sundarar went into the temple and prayed.

Sundarar reached the outskirts of Tiruvarur. He grieved that, due to his partial sight, he could not get a complete Darshan of the Lord. The very sight of the temple tower entranced him. Sundarar wanted to feast both his eyes on the beauty of the Lord and so prayed to Him to restore vision to his other eye also. Sundarar's supreme devotion and lamentation moved the Lord's heart. He at once cured the other eye also. Sundarar was extremely happy. He worshipped the Lord and remained completely absorbed in divine bliss.

In the mean time, people whom Paravayar had sent to greet Sundarar and inform him of her eagerness to meet him, found out that he had married Sangiliyar. They went back and told Paravayar about this. Paravayar was sunk in grief. She was annoyed, too. Paravayar's people refused to allow Sundarar's devotees to enter the house. When Sundarar heard of this, he was afflicted at heart. He sent some elderly devotees to Paravayar, to bring about a reunion. They failed. At midnight when all the devotees were asleep, he prayed to the Lord for His help in pacifying Paravayar. The Lord appeared before him and assured him of His help. The Lord disguised Himself as Sundarar's messenger, a Brahmin priest, and went to Paravayar's house. The Brahmin pleaded Sundarar's cause and asked her to accept him back. She refused, though she herself was grief-stricken at the separation from Sundarar. Her annoyance at his second marriage was so great! The Lord coolly returned to Sundarar, who was anxiously awaiting His return. When the Lord informed Sundarar of all that had happened, Sundarar fell down on the ground, in grief. 'If You do not help me, Oh Lord, I will give up my life.' The Lord seeing Sundarar's pitiable condition, assured him of help and again set out to go to Paravayar's house.

In the mean time, the devout Paravayar had understood that the Brahmin was no other than the Lord Himself and was suffering from terrible anguish for not recognising Him. The Lord again went to her house: and this time He appeared before her in His real form. Paravayar at once prostrated before Him. The Lord said: 'O Paravayar, I have again been sent by Sundarar to plead his case. Do not refuse this time. He is undergoing terrible agony on account of separation from you. Accept him and allow him to come to your house.' Paravayar prostrated to the Lord again, with folded palms and said: 'Oh Lord, first You came in the guise of a Brahmin but I did not recognise You.

Again You have come and have shown me Your real form. Oh Lord, how kind You are! You have graced my hut and showered Your grace on me. You have taken so much trouble this midnight, for the sake of Your friend. How can I go against Your wish? I will accept and obey Your command.' The Lord was immensely pleased with her. He blessed her and returned to Sundarar. The Lord informed him that he had pacified Paravayar and that he could now return to her. He then disappeared. Sundarar was overwhelmed with joy and sang the Lord's glories.

In the mean time, Paravayar had decorated her house beautifully and was eagerly waiting for her Lord. Sundarar, with his devotees, entered the house. Paravayar fell at his feet. Thus they were re-united after a long separation.

Yet, this was not all. The news that Sundarar had sent Lord Shiva Himself as his messenger, had reached the ears of Eyarkon Kalikama Nayanar of Tiru Perumangalam in Ponni Nadu. He was a great devotee of Lord Shiva. He was terribly angry with Sundarar for treating the Lord as a messenger to settle a domestic quarrel. How, in a wonderful and mysterious manner Lord Shiva brings about a reconciliation between the two Nayanars, both of them greatly devoted to Him, we shall see, when we describe the life of Kalikama Nayanar.

Equally interesting is the way in which the Lord unites in friendship Sundarar and Cheraman Perumal Nayanar, another royal devotee of the Lord. We shall describe it in detail when we come to the life of Cheraman Perumal Nayanar.

When, along with Cheraman Perumal, Sundarar was proceeding to Madurai, they arrived at Tirunagaikkoronam. Here Sundarar sang a song in which he asked the Lord to give him a pearl garland, precious stones, musk, spectacles, fragrance, clothes, jewelry, one-third of the wealth of Tiruvarur, horses which ran as fast as the wind, golden flowers, palanquin, etc. How wonderful is the relation between the devotee and the Lord! Sundarar regarded God as his friend, because God Himself had wanted it to be so. He adopted towards God the Sakhya Bhava (attitude of a dear friend).

In the company of Cheraman Perumal, Sundarar went on several pilgrimages, and met many of the kings of South India. Once, when they were at Tiru Kandiyur, they saw Tiruvaiyar on the opposite bank of the river. Cheraman desired to visit that place also. The river was in flood, and it was impossible to cross it. Sundarar sincerely prayed to the Lord to help them. He sang one of his songs which easily pleased the Lord. At once the river gave way, leaving a sandy tract through which they could walk across to the other bank. As soon as they reached the other bank, the river resumed its former form. They were delighted at this miracle of the Lord. Even the five elements are ever ready to serve the devotees of the Lord, at His command.

Cheraman then took Sundarar to his own place, with great honour and pomp. Sundarar stayed with Cheraman for some time. He suddenly remembered the Lord of Tiruvarur and wanted to go there. Cheraman could not accompany him and was therefore grief-stricken. Sundarar consoled him and asked him to stay behind and rule the country wisely and justly. Cheraman prostrated himself before him and gave him rich presents. He sent his own people to carry these presents, and to accompany Sundarar. It was the Lord's wish that Sundarar should receive gifts only from Him! How could His friend receive from others? Hence, He desired to deprive Sundarar of what Cheraman had given him.

When Sundarar and his retinue were passing through Tirumuruganpondi, the Lord sent His Servants to rob Sundarar of all the riches that he had received from Cheraman. The Lord's Servants disguised themselves as hunters and attacked the party that was carrying the presents. The party dropped all and fled. They went to Sundarar and reported to him what had happened. Sundarar went to the local temple and sang a Padigam. He came out of the temple and to his surprise he saw there all that they had been robbed of. Now, it was a gift from the Lord Himself, and so Sundarar, God's friend, could have it. Wonderful are the Lilas of the Lord.

After some time, Sundarar again desired to see Cheraman Perumal. On the way to Kundakolur, he went to Tiru Pukkoliyur Avinasi. As he entered this place, he heard simultaneously auspicious as well wailing sounds emerging from different houses.

On enquiry, he learnt how two Brahmin boys of the same age, from these two houses went to tank for a bath; and how one of them was caught by a crocodile, and the other escaped, providentially. The latter was being invested with the sacred thread that day, and hence the auspicious sound from that house. The people in the deceased boy's house were bemoaning the loss of the boy, sore at the feeling that, had he been alive, he would also be celebrating the sacred thread ceremony that day. Sundarar wanted to console the bereaved family.

As he stood in front of the house, the people stopped wailing and came out to receive Sundarar. They were eager for a long time to get his Darshan, and so, forgetting their sorrow, they came to welcome him. Their devotion moved Sundarar's heart. He was prompted by the Lord to bring the dead boy back to life. So, he went to the same tank and sang a song on the Lord of Avinasi to give the child back to the parents. The Creator, pleased with Sundarar, entered the stomach of the crocodile and re-constituted the body of the boy, though it had already been digested. Lord Yama, too, for his part, released from his custody, the life he had once removed and the crocodile vomited the boy! To the wonder of all, the boy was much more handsome than when he met with the accident, and he showed signs of growth, appropriate to the lapse of time.

All were amazed at this miracle of Sundarar. The parents of the boy were immensely pleased and embraced Sundarar's feet. Sundarar took the boy to the temple and worshipped Lord Avinasiappar. He himself performed the sacred thread ceremony for him.

At Kodunkolur, Cheraman, who had already come to know of the crocodile miracle, received Sundarar with still greater love and veneration than before. Sundarar stayed with the king for some time. One day Sundarar visited the temple alone and worshipped the Lord. The very sight of the Lord sent him into trance. He rolled on the ground shedding tears of God-love. The hairs on his body stood on end and his mind was filled with rapture. He regained consciousness after a long time. He was tired of worldly existence and so requested the Lord to take him back to Kailash. He sang a Padigam.

The Lord, desiring to take Sundarar back to His Abode, commanded the celestials to bring him to Kailash on a white elephant. He also informed Sundarar of this. Sundarar came out of the temple. The white elephant was waiting for him there. He mentally wished to take Cheraman Perumal also with him to Kailash. Then he climbed the elephant and proceeded towards Kailash.

In a Padigam he sang on this occasion, Sundarar himself reveals that this departure for Kailash was not in his physical body, but in his spiritual body. The physical body was discarded here in this world itself, and the elements of which it was composed were returned to their sources.

Cheraman learnt by intuition of Sundarar's departure for Kailash. At once, he mounted a horse and came to Tiru Anchaikalam. There he saw Sundarar going along the sky on the celestial elephant. At once Cheraman pronounced the

Panchakshara in the ears of the horse. The horse flew up and reached Sundarar. Cheraman worshipped Sundarar there. Both of them went to Kailash in their spiritual body.

At the Gate of Kailash, Sundarar was allowed to enter, while Cheraman was not. Sundarar went into His presence and praised His mercy: 'Oh Ocean of Mercy, You have pardoned my sins and released me from the quagmire of Samsara. You have taken me back into Your fold, and bestowed on me the Immortal Bliss. How kind and merciful You are!' He then informed the Lord that Cheraman was outside the Gate. To please His friend, Lord Shiva sent His Mount, Nandikesvarar to bring Cheraman also in. The Lord asked Cheraman how he could come to Kailash without His permission.

Cheraman replied that when he saw Sundarar proceeding to Kailash, he could not bear separation from him and so accompanied him. Now, by the good offices of Sundarar which earned for him the Lord's grace he had been admitted into the Lord's Abode. In these words, Cheraman expressed a very great truth: that even if the devotee is undeserving, if he is devoted to a saint (the Guru), he will also gain a place in the kingdom of God, through the intercession of the Guru.

Sundarar, as before, engaged himself in His service with all his heart and soul. Paravayar and Sangiliyar, being purged of their Karmas, also reached Kailash. They resumed their original duty as the servants of Mother Parvathi.

Tiru Neelakantha Nayanar: In Chidambaram, there once lived an ardent devotee of Lord Shiva. He was a potter by caste and profession. He had the highest regard for the devotees of Lord Shiva, too. He was ever eager to serve them. He was leading an ideal household life. He made beautiful begging bowls of clay and offered them free to the devotees of Lord Shiva, with great joy.

Shiva, in His aspect of Neelakantha was his sole refuge and prop. Hence, he was called Tiru Neelakantha Nayanar. He would always tell others how, for the protection of the world the Lord drank the virulent poison, and he would assure his friends that they who took refuge under His feet would be purged of all sins and would finally be taken to His Abode. In spite of his virtuous qualities, once he fell a victim to lust. One day, he visited the house of a prostitute.. When he returned home, his dutiful and pious wife understood this. This irritated her, though she did not show this and continued to serve him, as before. But, she had decided not to have any sexual relation with him. Nayanar could not understand the reason.

One day, as he approached her with passion, she took an oath and said: 'In the name of Neelakantha, I ask you: do not touch us.' Though she only meant herself, she had used the word us. Since she took the Name of the Lord and since she had used the word us, Neelakantha Nayanar decided that from that day he would not touch any woman in the world. Such was his sincere devotion to the Lord. They continued to live together. They did not want to make a fuss over their own resolve. No one knew about it. Years rolled by and they had grown old.

Lord Shiva wanted to reveal the greatness of His devotee and thus to immortalise his name. So, in the guise of a Shiva Yogi (a Saivite mendicant) the Lord came to Tiru Neelakantar's house. Neelakantar welcomed him and worshipped him. The Yogi gave him a begging bowl and said: 'Oh noble soul, kindly keep this in your safe custody, till I come back for it. To me it is extremely precious. It has the wonderful property of purifying anything that comes into contact with it. So, please protect it with the greatest care. Then the Shiva Yogi left the place and Neelakantar kept the bowl in a very safe place in the house.

After a long time, Lord Shiva came to the house of Neelakantar, as the same Shiva Yogi and asked for the bowl. The Lord Himself, by the power of His Maya, caused it to disappear from the house! Neelakantar searched for it, but could not find it. It was a mystery to him. He was ashamed of himself. Trembling with fear, he fell at the Yogi's feet and said that he could not find it. At this, the Yogi got very angry and accused Neelakantar, calling him a thief and cheat. Neelakantar offered to replace the bowl with a costlier one; but the Yogi would not accept.

Again and again Neelakantar pleaded that he had not stolen the bowl and that by a divine mystery it was missing from the house. The Yogi demanded that if that was the truth, Neelakantar should say so on oath, holding his wife's hand. When Nayanar, who had resolved, in the name of the Lord, not to touch anyone, declined this, the Yogi attributed this unwillingness to the fact that Neelakantar had in fact been guilty of theft. They went to the court. The Brahmins heard the case. They asked Neelakantar to promise, as desired by the Yogi. Neelakantar got into the tank, along with his wife; they had a stick in their hand, and each of them was holding one end of it.

The Yogi objected to this and wanted that Neelakantar should actually hold his wife's hand with his own. Neelakantar could not hide the secret relationship that existed between him and his wife any more, and so, related the whole story to the court. After this narration, Neelakantar and his wife caught hold of the two ends of the stick and took a dip in the tank. A miracle happened. As they emerged from the water, they shone with youth and beauty. The Shiva Yogi disappeared from their midst and Lord Shiva and Mother Parvathy appeared in the sky, blessing all of them. The Lord said: 'Due to the merit of having lived a life of self-control and devotion, you will live in My Eternal Abode, forever youthful.' The Lord thus revealed the glory of supreme devotion to Him (which alone made it possible for Neelakantar to refrain from lustful thoughts or actions, after his wife had sworn in the Name of the Lord) and a life of celibacy which bestows eternal youthfulness on you, and the unostentatiousness of a saint's virtue.

Iyarpahai Nayanar: 'Charity, free from the mean utterance I have none is found only among men of good birth' says the Kural. Among such noble souls Iyarpahai Nayanar ranked high. Charity was ingrained in him. It was his practice to invite Shiva Bhaktas to his house, worship them with faith and devotion and give them all they wanted. He had taken a vow never to say no to what a Shiva Bhakta wanted.

Iyarpahaiar was a native of Kaveripoompattinam. He was a Vaisya by caste. To him Shiva Bhaktas were the living manifestations of Lord Shiva. Lord Shiva was pleased with His devotee. He wanted to reveal his true greatness to the world. So, the Lord, in the disguise of a Brahmin, with sacred ashes smeared all over his body, came to Nayanar's house. He welcomed the Brahmin with great joy, as the very sight of the holy man thrilled the Nayanar.

The Brahmin said: 'Oh noble soul, you are far famed for your charitable nature. Learning that none returns empty-handed from your house, I have come to you for a gift. I shall disclose it to you, if you promise to give what I want.' The Nayanar agreed readily 'provided I have it with me.' The Brahmin at once revealed what he wanted: 'It is the gift of your wife.' Nayanar had no difficulty at all in granting this! The supreme devotee of the Lord that he was, he did not stoop to doubt the credentials of the Brahmin who, though he appeared to be a Shiva Bhakta, had such an undesirable desire: such is the unquestioning nature of devotion. Nor would Nayanar hesitate to fulfil the Bhakta's wish, on the plea that it involved unrighteousness: for, to him worship of the guest (Guest is God) was greater Law than all the moral codes.

Nayanar went inside the house and informed his wife of all that had happened. She was shocked at first, but quickly regained her composure. To a chaste wife, the husband is God, and whatever he commands is Law and Dharma. She readily agreed to follow the Brahmin-guest as his wife. Nayanar came out with his wife and asked the Brahmin to accept the gift. The Brahmin, however, feared the wrath of the wife's relatives and asked Nayanar to accompany them till they were safely out of the town and out of danger. Nayanar agreed to do so and armed himself to protect the Brahmin. They then proceeded to go.

In the meantime the relatives of Nayanar's wife came to know of the whole story and were furious. They followed the Nayanar and party and threatened the Brahmin with death, unless he abandoned his impious desire. The Brahmin pretended to be scared. Nayanar's wife, however, assured him that Nayanar was capable of defeating them all. Nayanar was ready to fight them. The relatives endeavoured to convince Nayanar of the unrighteousness of the whole thing, and, when they found that they could not, they preferred to die at his hands, than submit to the shame. Nayanar at once pounced upon them and chopped off their heads. All of them died and Nayanar, happy at the thought, that through the grace of Lord Shiva, he had succeeded in keeping his vow of worshipping His devotee, proceeded further with the Brahmin and the wife. When they reached the temple of Tiruchaikadu, the Brahmin asked Nayanar to leave them and return. Nayanar prostrated to the Brahmin and turned his steps homeward.

As he had hardly proceeded a few yards on his homeward journey, the Brahmin again called Nayanar aloud. Thinking that there might have been another attack on the party, Nayanar hastened to where the Brahmin was: but, to his amazement, found that he had disappeared and that his wife was standing alone there. He searched here and there for the Brahmin, and was worried when he could not be found. Lo and behold, Lord Shiva and Mother Parvathy appeared in the sky and blessed Nayanar and his wife: 'Oh noble souls, I am immensely pleased with your devotion to My Bhaktas. Both of you will very soon reach My Abode.'

With these words, the Lord disappeared. The Nayanar and his wife reached His Abode and rejoiced there. Nayanar's relatives who died at his hands also attained the lotus feet of the Lord.

Thus had the Lord proved the nature of supreme devotion, which does not question. And, the Lord also revealed the truth that such unquestioning devotion does not result in the violation of the Dharma. All glory to the Lord and His devotees.

Ilayankudi Mara Nayanar: Ilayankudi Mara Nayanar was a farmer. Maranar was his name: he lived in a village called Ilayankudi. He was a great devotee of Lord Shiva and His devotees. He took the greatest pleasure in serving them. This Sadhana was known as Maheswara Puja (or worship of the devotees of Maheshvara or Lord Shiva). It is described as follows:

'On seeing a devotee of Lord Shiva, with the external marks of Vibhuti, Rudraksham, etc., taking him as

Lord Shiva Himself, welcoming him, prostrating before him, washing his hands and feet, drinking that water (Charanamrit), giving him a seat, worshipping him with flowers, Dhoopa, Deepa, and Naivedya, pleasing him with sweet words, thanking the Lord for the opportunity, and accompanying the guest for some distance while sending him away—these constitute Maheshwara Puja. This is also included in Chariyai (one of the main Saivite disciplines).'

Offering food to the Lord's devotees had purified his heart and made him a fit receptacle for the grace of God. As Tiruvalluvar has said in the Kural:

Fortune dwells with a delighted heart in the house of the man who honours his guest with a pleasant countenance.

Nayanar had been blessed with all the wealth of the world. But, he considered that the wealth belonged to the Lord, to be utilised for the benefit of His devotees.

Lord Shiva was highly pleased with the Nayanar's devotion. He wanted to show to the world that His devotee would be undaunted by the worst calamity and would remain unshaken in his virtue. Nayanar's wealth melted away. His wealth had left him, but not his virtue. On the contrary, his devotion to the Lord and His devotees grew more and more intense. Nayanar sold all his property and had to sell even himself in order to be able to serve the devotees of the Lord.

One day it was raining heavily. Nayanar and his wife were starving. No one came forward to help them. Finally, he bolted the door and was about to fall asleep. Just then he heard a knock at the door, and, on opening it, found a Shiva Bhakta standing in front of the house, drenched with rain. Nayanar at once took the guest inside, dried his body and gave him fresh clothes to wear. 'Rest awhile, Swami, while we prepare some food for you to appease your hunger,' said Nayanar and told his wife of his predicament: there was nothing to offer the devotee of the Lord. But, the devout wife suggested that Nayanar could go into the backyard and collect the grain-seeds that they had just sown that very day. Nayanar accepted the suggestion. On account of the heavy rain, the grains were floating and it was easy to collect them in a basket. As soon as he brought the grains, the wife fried and crushed them, and with the help of some greens that grew in their own backyard, cooked a nice dinner for the guest.

Nayanar was supremely happy. And, as he went to awaken the guest, he discovered that he had disappeared. At the same time, Nayanar saw in sky, Lord Shiva Who had come in the form of the devotee and Mother Parvathy showering Their blessing on him and his wife. The Lord said: 'Oh noble souls, I am highly pleased with your devotion. You will soon attain My Abode and live there for ever.'

Viralminda Nayanar: Viralminda Nayanar was born in Sengunru, a hilly place. He was a Vellala by caste.

He was a staunch devotee of Lord Shiva. Through His grace, he was free from 'I'-ness and 'mine'-ness. He had equal vision. He served His devotees and attained purity of mind. To him worship of Shiva Bhaktas was equal, if not even superior to the worship of Lord Shiva Himself. He felt that no one could get Shiva's grace without first worshipping Shiva Bhaktas, and that he who worships even the Shiva Lingam with all faith and devotion, would not attain salvation if he insults Shiva Bhaktas. Daily he used to visit the temple. Before worshipping the Lord, he used to worship the Shiva Bhaktas who might be found there.

He left Sengunru on a pilgrimage and came to Tiruvarur. One day when he was worshipping the Lord, Sundaramurthi Nayanar came to the temple. Sundarar by-passed the Bhaktas who were in temple and went into the sanctum sanctorum to worship the Lord. This upset Viralmindar, who was observing this. He could not tolerate this insult to His Bhaktas. He said to Sundarar: 'You have insulted the Shiva Bhaktas. By this act you have rendered yourself unfit to remain in the holy circle of Shiva Bhaktas. Hence, you are excommunicated from this circle.' He added further: 'And, Shiva, for having so thoughtlessly accepted such improper worship at your hands, He, too, shall be regarded as an outcaste from the divine fold.' So firm was he in his conviction that he could thus 'reprimand' God Himself! In fact, it was Shiva Himself Who spoke through him to instruct His Bhaktas in the proper attitude they should have towards His Bhaktas.

Sundarar immediately understood Viralmindar's inner Bhav towards the Bhaktas as well as towards Lord Shiva, and prostrated before him. He then sang a Padigam praising him. The Padigam melted Viralmindar's heart so much that he greeted Sundarar and said: 'Your mind is well established in the service of Shiva Bhaktas. You have got sincere devotion to them.' Lord Shiva was greatly pleased with Viralmindar's great steadfastness in his devotion to Shiva Bhaktas. Thus had the Lord revealed the great glory of the Bhakta. He was then elevated to the blessed plane of the Shiva Ganas where the Lord made him leader of the Ganas. Glory to such Bhaktas!

Amaraneedi Nayanar: Amaraneedi Nayanar was a Vaisya by caste. He belonged to Pazhaiyaarai in the Chola Kingdom. Pazhaiyaarai was a very fertile place, surrounded on all sides by gardens and green fields. In those days this place was very famous.

Amaraneedi Nayanar was a trader in gold, diamonds, silks and cotton goods. He used to import these goods from foreign countries and was selling them at reasonable prices. He earned money honestly and became rich. Though he was engaged in worldly activities, his mind was fixed on Lord Shiva. He was an ardent Shiva Bhakta. He would invite Shiva Bhaktas to his house and worship them. He would give the Kowpeenam, cloth, etc., and feed them nicely and send them away happy, with other gifts.

He used to visit the sacred temple of Tirunallur during festivals and worship Lord Shiva with intense faith and repeat Panchakshara Mantra daily with Bhava. Not being satisfied with this visit during festivals only, he wanted to settle down there once for all, always enjoying the Lord's Darshan, and feeding Shiva Bhaktas. So he left Pazhaiyaarai and migrated with his family and relatives, to Tirunallur. He built a beautiful

Mutt there to accommodate Shiva Bhaktas who visited the temple. Daily he used to invite Shiva Bhaktas and offer Kowpeenam, etc.

Lord Shiva was highly pleased with Amaraneedi Nayanar's Kowpeena charity and extreme kindness to Shiva Bhaktas. He wanted to show to the world His Bhakta's greatness and also shower His blessings on him.

So, one day Lord Shiva in the guise of a Brahmachari, with beautiful matted locks on his head, sacred ashes on his forehead, with a staff on his shoulder, appeared before Amaraneediar's Mutt. Two Kowpeenams and a small ash-bag were tied to one end of the staff. He had a charming face. His eyes were glittering. He walked gracefully into the Mutt. Amaraneediar, with extreme joy, welcomed him and worshipped him.

The Brahmachari said: 'Oh friend, you are a noble soul. People are highly praising your Kowpeena charity. I have come to you for Darshan.' Amaraneediar begged of him to take Bhiksha. He readily agreed and said: 'I shall go to the river and return after finishing my bath and Nitya Karmas. Rain may drench my Kowpeenams. So, please keep this dry Kowpeenam safely with you, and I shall come back for it. The Kowpeenam is very precious, as you already know. So, please keep it safe.'

The Brahmachari went away, and Amaraneediar kept the Kowpeenam safely inside the house. But, the Lord willed that it should disappear! Soon after the Brahmachari came back after his bath, etc., and asked for the dry Kowpeenam as rain had drenched the Kowpeenam he had on the staff. Amaraneediar could not find it. He prayed hard to the Lord. Yet, he could not find it. He approached the Brahmachari, trembling, with another Kowpeenam, and explained his predicament to him.

But, the Brahmachari was in no mood to take any explanation. Amaraneediar offered much wealth, etc., in compensation. But, the Brahmachari said: 'What have I to do with all this wealth? All these are of no use to me. I only need a Kowpeenam.' And, in saying so, the Lord in the guise of the Brahmachari, uttered a very great truth. He continued: 'I have got another Kowpeenam: you can give me another of the same weight.' Amaraneediar was greatly relieved when he heard this. He brought a balance. He put the Kowpeenam on one side and another piece on the other. The Brahmachari's scale went down. Whatever Amaraneediar put on his side, the Brahmachari's scale was heavier. Amaraneediar was amazed: and he understood that it was God's own Lila. All his wealth could not equal the Brahmachari's Kowpeenam! How could it? Lord Shiva's Kowpeenam represents the Vedas. The fibres of His Kowpeenam represent the Shashtras.

Amaraneediar was on the horns of a dilemma. He fell at the Brahmachari's feet and asked him to allow himself, his wife, and his child to be weighed against the Brahmachari's Kowpeenam. The Brahmachari agreed. Amaraneediar got on the scale with his wife and his child, saying: 'If I have truly served the Shiva Bhaktas, with faith and sincerity, let this scale be equal in weight to the other one.' Immediately the two scales were equal. The merit of Amaraneediar's selfless service of the Shiva Bhaktas was equal to the merit of Lord Shiva's Kowpeenam.

The people who witnessed this were wonderstruck. They prostrated before Amaraneediar and praised him. Devas from the heaven showered Parijatha flowers. The Brahmachari disappeared and Lord Parameswara and Mother Parvathi appeared on Their Rishabha before Amaraneediar, his wife and child. He blessed them: 'I am immensely pleased with your whole-hearted and sincere service of My Bhaktas. I am immensely pleased with your Kowpeena charity. You three will come to My Abode and live there happily for ever.' On account of the Lord's grace, the balance itself turned into a celestial car in which Amaraneediar, his wife and his child attained Shiva's Abode.

Eripatha Nayanar: Eripatha Nayanar was born in Karuvur, one of the main cities of the Chola Kingdom. It was a very sacred place, situated on the bank of the river Ambiravati. On both the banks of this river saints and sages were doing Tapas and were radiating spiritual vibrations. A famous temple was there, too, dedicated to Lord Pasupatheesvarar Who was showering His grace on the king and the people alike. They were all happy. Eripatha Nayanar was daily worshipping Lord Pasupatheesvarar with great faith and devotion. His one aim in life was to serve Shiva Bhaktas and to offer them every kind of protection. He always carried a weapon, an axe, for this purpose. With the axe he would punish anyone trying to harm Shiva Bhaktas. He was doing by this the Lord's own work!

In that city, there lived a Shiva Bhakta by name Shivakami Andar. He was very regular in his daily worship of Lord Shiva. Early morning would find him in the garden after bath, collecting flowers, making garlands for taking to the temple and offering to the Lord. This was his routine.

On a Maha Navami day when all the people were jubilant, Shivakami Andar was rushing to the temple, as usual, with a basket of flowers. At the same time, the king's pet elephant was returning from the river, after its bath. On its back were two Mahouts, and three others were escorting it. Suddenly, it went mad and was chasing the people. They were running here and there. It ran towards Shivakami Andar. It caught hold of him, wrenched the basket of flowers from him, threw it on the ground and ran away.

The flowers were all scattered on the ground. Shivakami Andar was greatly upset. The elephant had destroyed the flowers he had kept for the worship of the Lord. He chased the elephant. He was very aged and soon fell down exhausted. He was weeping bitterly, crying aloud: 'Shivada, Shivada' (a cry expressing agony). Eripatha Nayanar happened to pass that way. He heard Shivakami Andar's pitiable cry and the cause of it. 'Where is that elephant?' asked Eripathar and began to run in the direction indicated by Shivakami Andar. Soon he overtook the elephant and hurled his powerful axe, killing it with one stroke. Then he pounced on the Mahouts and killed them, too.

The news of the elephant's fate reached the king who immediately reached the spot on his horse, surrounded by his soldiers. He could not see who had killed the elephant, for, he could not associate the Shiva Yogi Eripathar with such an act. He began to shout: 'Who killed my elephant?' When someone pointed to Eripathar, immediately the king's wrath vanished, for he knew that if the Shiva Yogi had done so, there should have been a very valid reason for it. 'He must have killed it in self-defence,' thought the king and felt happy that the elephant had done no harm to the Shiva Yogi. He addressed Eripathar: 'Oh Swamin, I did not know that you killed the elephant. Definitely, the elephant and the Mahouts must have done some harm to you and you rightly punished them.' Eripathar narrated to the king all that had happened, and said: 'Since the elephant and the Mahouts were guilty of Shiva-Aparadham, I killed them.'

The moment the king heard the expression Shiva-Aparadhara (sin against Lord Shiva) he suffered terrible mental agony. He fell at the feet of Eripathar and said: 'O Swamin, for what they have done, the punishment awarded by you is not enough. I have committed a great crime by keeping such an elephant and such Mahouts. Now, I do not deserve a death through your holy weapon, the axe. Here is my own sword. Please be gracious enough to cut off my head with it.'

Eripathar was stunned to hear these words. He himself was struck with remorse. 'What a great pain have I inflicted on the king! What a noble king he is!' he thought; and, lest the king should execute the punishment on himself, he took the sword from the king. Eripathar felt that he was the cause for the king's affliction, and in self-punishment, he began to cut his own throat. The king was alarmed. He thought that he would now be guilty of another offence and at once gripped the sword and stopped Eripathar from cutting his own throat.

The Lord's Lila was over. A voice was heard in the sky: 'Oh noble souls! This is Lord Pasupatheesvarar's Lila. It is His wish that His Bhakta's sincere and faithful service to Him must be recognised by the world.' Immediately, the elephants and the Mahouts got up, as from sleep. Shivakami Andar's flower basket was full. All were amazed and began to sing Lord Pasupatheesvarar's glory. Eripathar placed the sword at the king's feet and prostrated to him. The king also fell at Eripathar's feet. Both embraced each other and were in great joy. Eripathar wished that the king should mount his pet elephant. The king did so. Eripathar returned to his place. Shivakami Andar went to the temple with the flowers. Eripathar continued to serve Shiva Bhaktas. Finally he cast off his mortal coil and reached the Abode of Lord Shiva.

Enadinatha Nayanar: Enadinatha Nayanar was a Shanar (toddy tapper). He was born in Eyinanur in Chola Kingdom. It was situated to the Southeast of Kumbakonam on the bank of the river Arisol. It was very fertile and rich.

Enadinathar was an ardent devotee of Lord Shiva. Like Maiporul Nayanar, however, he was devoted even to the external marks of Shiva Bhakti. To Enadiar, the three white lines of Vibhuti or sacred ash on one's forehead were sufficient to evoke his own reverence.

It would not be out of place here to say a word about this mark on the forehead of devotees of Shiva. Through this mark Lord Shiva teaches silently that the spiritual aspirant should destroy the three types of impurities, *viz.*, Anavam (I-ness), Karma (selfish activity), and Maya (illusion): the three desires or Eshanas, *viz.*, desire for worldly goods, for son and for wife: the three Vasanas or subtle tendencies, *viz.*, Lokavasana (worldliness), Dehavasana (attachment to the body) and Shastravasana (blind faith in the scriptures and polemics), and that he should transcend the three bodies (physical, astral and causal), and the three states, *viz.*, waking, dreaming and deep sleep,—and eventually attain union with Lord. The Shashtras assure us that the Bhasma or sacred ash is a divine healer. It cures all diseases, including the disease of birth and death, and bestows on the devotee who wears it, the highest wealth, *viz.*, Moksha.

Such is the glory of the sacred ash: and, no wonder Enadiar worshipped whoever came to him with the ash on his forehead. Enadiar saw Lord Shiva in him. He was ready to give even his own life at the feet of the devotee who wore the ash. Enadinatha Nayanar was a very good swordsman. He was a tutor to the princes in fencing. He earned a good income from his profession. He spent all his income in the service of the Shiva Bhaktas. He became very popular, too. This evoked the jealousy of another man belonging to the same profession, by name Atisuran. Contrary to his name (which means a great hero), he was not at all skilful and not strong either, because he was full of vices. Yet, he wanted to fight with Enadinathar and defeat him.

One day Atisuran marched towards Enadinathar's house, with all his relatives, fully armed: he hoped to defeat Enadinathar, with the help of his relatives. He stood in front of Enadinathar's house and challenged him to a fight—jackal coming to fight the lion. Enadinathar accepted the challenge and came forward to fight. Atisuran got frightened. He asked Enadinathar to come to the nearby grove to fight with him. The relatives of Atisuran were waiting in the grove. In the mean time, the friends of Enadinathar had also gathered around him. The two parties fell on each other, and in the terrible fight that ensued many lives were lost. Atisuran ran away from the grove. He wanted to kill Enadinathar, not in open fight (which was impossible), but by strategem.

The next day, he sent a message to Enadinathar: 'Let us fight again, but without any assistance this time: otherwise, many innocent people die on our account. Let us go to a lonely place, without anyone's knowledge and fight.' Enadinathar accepted it. The next morning, Enadinathar went away secretly and was awaiting Atisuran's arrival at the stipulated place. Atisuran, with the sacred ashes on his forehead (which was cleverly hidden by his shield) approached Enadinathar. Enadinathar pounced upon him, with a big roar.

In a moment, Atisuran removed the shield, revealing the sacred ashes. Enadinathar quickly lowered his sword and thought: 'What a sin I was about to commit! He has become a Shiva Bhakta now. I must not harm him. Let him achieve his object of killing me.' Endinathar wanted to throw the sword away, but kept it in his hand, else he would be compelling his opponent (a Shiva Bhakta!) to incur the sin of killing an unarmed person. As he was mutely standing thus, Atisuran killed him.

Lord Shiva was highly pleased with this self-sacrificing devotion that Enadinathar had for the ashes. He appeared before Enadinathar as he fell, and took him to His Abode.

Kannappa Nayanar: Nagan was the king of hunters at Uduppur in Pottapi Nadu. His wife was Tattai. They were great devotees of Lord Subramanya. By His grace, they had a child, after a long time. It was very heavy: so, they named him Tinnanar.

Tinnanar was Arjuna in the previous birth, according to Tiru Kalahasthi Puranam. When he went to worship Shiva, to get Pasupatha Astra, and when the Lord came to him as a hunter, Arjuna did not recognise Him. So, he had to be born as a hunter again and adore the Lord, before attaining Final Liberation.

Tinnanar was educated according to the hunters' customs. He became a good archer. Even when he was young, his father retired, and crowned him king. Though he was a hunter and carried on hunting as his Dharma, Tinnanar was full of love and would not kill young ones, females, diseased animals, etc. Spiritually, he had already killed the animals within himself, *viz.*, lust, anger, greed, vanity, etc.

One day, Tinnanar went out hunting. A pig escaped from its net and was running away. Tinnanar pursued it accompanied by two others, Nanan and Kadan. The pig was tired and stood near a tree. It was quickly killed by Tinnanar. They were tired, too, and thirsty. They proceeded towards the Ponmukali. Tinnanar wanted to climb the nearby mountain. Nanan, too, volunteered to follow him, saying that on that, the Kalahasthi hill, there was Lord Kudumithevar (God with a Tuft). Kadan was busy cooking the pork.

Even when he began to climb the hill, there was a definite change coming over Tinnanar, owing to past Samskaras. He felt that a great burden was being lifted off his shoulders. He was losing body-consciousness. As he saw the Lord there, he felt supreme love surging in his heart. He embraced the Lingam and kissed It. He began to shed tears of joy. He felt that the Lord was lonely there, and that he should thenceforth remain with Him. Again, he thought that the Lord might be hungry. Though he was reluctant to leave the Lord alone, he quickly came down the hill to fetch some food for the Lord. He took the best pieces of the pork, tasted them and ear-marked the very best for Him. In the mean time, he gathered from Nanan that the Lord was worshipped daily with water, flowers, etc, before the food was offered to Him. So, he began to collect the other articles of worship. He filled his own mouth with water from the river. Flowers, he gathered and wore them on his head! He took the pork, bow and arrow and went up the hill again, alone this time.

At the temple, Tinnanar poured from his mouth, the water that he had brought for His worship. That was his 'Abhishekam'. Then he decorated the Lingam with the flowers he had brought on his own head. This was his 'Archana'. He then placed the pork before the Lord. He went out and stood guard for Him, at the entrance, lest some wild animals should hurt Him. In the morning again he went out to hunt and bring fresh food for the Lord.

In the mean time, Nanan and Kadan worried about the change that had come over Tinnanar (which they thought to be madness). They went and reported the matter to Tinnanar's parents. They came and tried, in vain, to take him back. They, too, went away.

When Tinnanar left the temple in the morning to get food for the Lord, Shivagochariar, the temple priest, came there for the usual orthodox worship. He was horrified at the desecration that some unknown person had done in the temple. He was well versed in the Agamas (rituals of Shiva-worship). He performed the necessary purificatory rites and took bath again and began his formal worship. He brought water in a holy pot, with a bandage around his own mouth, lest the breath of his mouth should pollute it. He brought fresh flowers in a holy basket. He brought fruits and sweets, newly made and unpolluted by anyone tasting it, before the Lord for being offered to Him. He went home after the worship.

Tinnanar returned with fresh meat. He removed the priest's decorations, and did the worship in his own way, and then as usual, stood guard at the entrance.

This went on for five days. The priest was greatly upset about the desecration of the holy place. He appealed to the Lord to stop it. Lord Shiva wanted to show to Shivagochariar the nature of Tinnanar's supreme devotion. He commanded him in a dream, to hide himself behind the Lingam, when Tinnanar went to the temple the next day, and watch what took place.

On the sixth day, Tinnanar went out as usual for getting the Lord's food. Whiie returning, he saw many ill omens, which made him feel that something had happened to the Lord: he was so unconscious of himself, that he did not think that something could happen to him. He ran towards the Lord. He was grieved to see blood issuing from the Lord's right eye. The articles he had brought for the worship dropped from his hand. He wept bitterly. He could not find who had done this to the Lord. He treated the eye with herbs he knew of. Still the bleeding did not stop.

A simple idea occurred to him: 'flesh for flesh'. At once, with his own arrow, he took out his own right eye, and fixed it over the right eye of the Lord. The bleeding stopped. He was very happy. When he was dancing in ecstasy, he noticed that the Lord's left eye had begun to bleed. But, he had already found out the remedy. There was only one problem: how to locate the eye of the Lord, when his own eye had been pulled out. So, Tinnanar planted his foot at the place where the Lord's left eye was on the Lingam, and began to pull his left eye out, with his arrow.

At once, Lord Shiva caught hold of his hand and said: 'My dear child, Kannappa! Stop plucking your eye.' The Lord repeated the word Kannappa thrice. Kannappar was thrice blessed. Tinnanar became Kannappar, because he gave his own eye to the Lord. Lord Shiva took him with both Hands, and kept him on His right side. Kannappar regained his vision and lived as god himself. Shivagochariar understood the true nature of devotion.

This story has an esoteric meaning, too. Nayanar had conquered all other evils: but, Anava Malam or egoism had to be killed, too. The wild pig represents this. Supreme Bhakti dawned, the moment this was killed. In its chase, the seeker is accompanied by good and evil (the two hunters Nanan and

Kadan). Nanan (good) described the glory of the Lord to him: Nanan represents good Samskaras. Kadan (the evil) had to be left behind. The aspirant with good Samskaras, goes to His Presence. But, when he has to attain God-realisation, even this has to be renounced. Hence, Nayanar, when he went to worship Him, went alone. Nayanar's parents (the hidden good and evil tendencies and worldly desires) tried but failed to take him away from God. The Lord asked the priest to hide behind Him, while Tinnanar was in front: this means, true Bhakti is far superior to mere ritual. Tinnanar's readiness to pluck out his own eyes for His sake is total self-surrender or Atma-Nivedan, the highest peak of devotion which immediately reveals the Lord in all His glory.

Kungiliya Kalaya Nayanar: Kungiliya Kalaya Nayanar was born in Tirukadavur in the Chola kingdom. The Lord of this place is called Amirda Ghateswarar. Once Devas and Ashuras came to this place with nectar in a pot. They wanted to take bath. So, they left the pot on the ground and went to the river. When they came back to the place, they could not lift the pot. The pot itself had been transformed into a Lingam. Hence this Lingam is known as Amrita Lingam. Markandeya worshipped this Lingam and became an immortal boy of 16 years.

The Goddess in this place is called Abhirami Amman. Abhirami Pattar, a great devotee of Mother, sang beautiful songs in praise of Her: and the Mother Who was highly pleased with this, changed the new moon day into a full moon day, in order to save him from the king's wrath.

Kungiliya Kalaya Nayanar was a Brahmin by caste He got the name because he was always holding a pot (an incense pot) in his hand. He considered burning of incense before the Lord was the best service to Him. Lord Shiva was highly pleased with the Nayanar's intense devotion and his wonderful service. He wanted to put it to test, so that the true glory of his supreme devotion to the Lord may be understood by all.

By the will of Lord Shiva, Nayanar became poor suddenly. He sold all his property. His family was starving. Still, he continued to burn incense before the Lord. One day his dutiful wife thought: 'Everything has been sold. Only this Mangalyam (a sacred thread with a pendant, which every married woman must always have on her person, till the husband dies, when it is removed), is left. I will give it to my Lord: though it is inauspicious to do so.

Let him sell it and obtain some rice, with which we could feed the children who may die of hunger otherwise.' She removed the Mangalyam and gave it to her husband, who gladly received it. As he was proceeding to the market to sell it, Lord Shiva Himself appeared before him, in the guise of a hawker and said that he had very good incense. The word incense at once made Nayanar forget himself and the mission! He quickly bought incense for the price of the Mangalyam, and went to the temple to burn it before the Lord.

His wife patiently waited for his return, and, not finding him even after nightfall, put the children to bed and remained praying. The Lord was immensely pleased with this noble couple. The faithful wife was prepared to part with even the most sacred ornament for the service of her lord, her husband. The Kural says: 'Rain falls at the bidding of her who, on waking from sleep, worships no other God but her husband.' That night Lord Shiva appeared in her dream and blessed her with all wealth.

She woke up from her sleep and was amazed to find all types of wealth in the house. She sang His glories. Immediately she prepared a nice meal and was waiting for her lord's return.

After blessing the Nayanar's wife, thus, Lord Shiva appeared before Nayanar in the temple and said: 'Oh noble soul, I am immensely pleased with your devotion. Your dutiful wife is anxiously waiting for you in the house with milk and food. Kindly go to your house.' It was only then that Nayanar became aware of this world! He returned to the house and found that it had been transformed into a heaven, by the grace of the Lord. Shiva Bhaktas, too, had assembled in the house in large numbers. They all sang the glories of the Lord. The Nayanar treated the wealth that the Lord had bestowed upon him as the property of Shiva Bhaktas and served them.

One day Nayanar wanted to visit the temple at Tiruppanandal. The Lord of this temple is Arunasatesar. Thatakai was the daughter of an Asura. For getting a son, she worshipped the Shiva Lingam regularly. One day at the end of the worship, she wanted to garland the Lingam. As she lifted the garland with both her hands, her cloth began to slip from her waist. She held it with her elbows, and hence could not raise her hands (and the garland) high enough.

To relieve her, the Lord leaned to one side and accepted the garland. Many people tried to pull the Lingam straight: but it could not be done. Nayanar heard that the king of the place was upset about it and wanted the Lingam to be straightened. Nayanar wanted to help the king. He tied the Lingam to his neck with a rope (the rope of God-love) and gently pulled it. The Lingam became upright! Devas rained flowers from heaven. All were amazed and recognised the glory of the Nayanar and his great devotion to the Lord.

After spending some more time in the service of Lord Shiva and His Bhaktas, Nayanar reached His Abode.

Manakanchara Nayanar: Kancharur was a fertile place in the Chola kingdom. The people were all Shiva Bhaktas. In this place there lived a staunch devotee of Lord Shiva by name Manakancharanar. He was a Vellala by caste. He was a hereditary Senathipathi. People of the community had the highest regard for him. He was a contemporary of Sundaramurthi Nayanar. To him adoration of Shiva Bhaktas was the highest form of worship of the Lord. He would read their minds from their look, and would serve them without their asking.

He had no children for a long time. He worshipped Shiva with faith and devotion and obtained the boon of a daughter from Him. Nayanar celebrated the birth of this divine child, with a lot of charity. In due time, the girl attained the marriageable age. She was engaged to be married to Eyarkon Kalikamar who was also an earnest and sincere devotee of the Lord. The date of the wedding had been fixed and all arrangements made.

In the mean time, Lord Shiva wanted to shower His supreme grace on the Nayanar. He took the form of a Maha-Vrathiar (man of great vow) who wears the sacred ash on his forehead, matted locks adorned with a garland of bones, and a sacred thread made of human hair on his chest. The Maha-Vrathiar appeared before Manakancharanar who received him with great delight. When the ascetic enquired about the cause of the festive appearance of the house, Nayanar explained that his daughter was to wed that clay. He asked the girl to bow to the ascetic and receive his blessings.

The ascetic saw her flowing hair, and said: 'Oh noble soul, I am delighted to see her hair. This can be conveniently made into a Panchavati (the thread that adorns my chest).' At once, Nayanar took a knife and, without thinking for a moment, cut the hair on his daughter's head and handed it to the ascetic. In his extreme devotion to the Shiva Yogi, he did not even consider the fact that he was disfiguring his only daughter, and that the bridegroom might refuse to accept her. The Lord in the form of the ascetic immediately disappeared. He gave Nayanar and his family Darshanam along with Mother Parvathy and blessed them.

Eyarkon Kalikamar, the bridegroom, and his party arrived there soon after, and came to know of all that had happened. He was sorry that he had not come earlier and had the Lord's Darshan. When he saw the disfigured bride and hesitated to accept her, Lord Shiva, the Indweller, understood the cause, and restored the hair to her head. Nayanar and his family were very happy and proceeded with the wedding.

Arivattaya Nayanar: There once lived in Kannamangalam in the Chola kingdom a rich Vellala by name Thayanar. He was leading the life of an ideal Grihastha (householder) of whom the saint Tiruvalluvar has sung:

> *He will be placed among the gods in heaven who in this world follows the law of the householder's life.*

Thayanar was a great devotee of Lord Shiva. His devotion took the form of a daily offering to the Lord of food prepared with red rice, a sauce made of red herb, and mango pickle. He considered this as an act of great devotion to the Lord. The Lord was highly pleased with Thayanar's devotion. He wanted to put it to the test, in order to manifest it to the world in all its glory. By His Will, poverty struck Thayanar. Thayanar got himself employed and earned his wages in kind (red rice). He himself would not eat this red rice, but lived on the inferior khar rice.

The Lord tested him further. All the fields in the place grew only red rice. But, Thayanar would not touch it. His wife cooked for him some green leaves from their garden. Thayanar was content and was intent on his usual offering to the Lord. The Lord put His devotee through more severe tests. Even the green leaves withered away and there was nothing left. Thayanar was not at all perturbed. He happily lived on mere water: his mind was full of the bliss of the worship of the Lord and he felt neither hunger nor thirst. One day, Thayanar, now emaciated and weak, was taking his usual offering to the Lord, followed by his wife. He stumbled on the way and fell down.

The offering he had, was spilt on the ground. Thayanar was greatly upset. He began to weep bitterly: 'Oh Lord, today the food intended for You has been spilt on the ground. What great sin have I committed to deserve this? Please forgive me. Have mercy on this poor creature. You are omnipotent, omniscient and omnipresent. If this is true, You must be present here also. Kindly come and accept the offering here. If You do not eat this, I will give up my life.' With these words, he began to cut his throat with an Arival (sickle). Hence, the name Arivattaya Nayanar.

Lord Shiva was highly pleased with his devotion. His Lila was over. He at once stretched His hand and caught hold of Nayanar's, thus preventing him from cutting his own throat. Nayanar did not realise what was happening. At that time, he heard the sound of someone biting a mango pickle. He understood the Lord's Lila, sang His glories and danced. The Lord and Mother Parvathi appeared before him and blessed him: 'Oh noble soul, I am immensely pleased with your devotion. You and your chaste wife will soon come to My Abode and live happily there.'

Anaya Nayanar: Tirumangalam was an important place of pilgrimage in Mazhanad (Trichnopoly District). The Lord Who dwells in this place is called Samavedesvarar. It was He Who purified Parasurama of the sin of killing his mother and also gave him the axe.

In that enchanting place, there was a cowherd by name Anayar. Because he was tending cows, he was known by that name. He was a staunch devotee of the Lord. He was devoted to the Bhasma, and also to Shiva Bhaktas, irrespective of their caste. His devotion to the Lord took the form of playing on his flute the holy Panchakshara Mantra of Shiva. He aspired to realise the Lord, through this Mantra.

One day when he was playing the Mantra on his celestial instrument, under a Konrai tree (a favourite of Lord Shiva), the music captivated all the cows and calves. Even the birds sat on trees and silently heard the enrapturing music of the flute. The peacocks danced in joy, keeping time with the music. Hearing the music, other animals stood motionless. The music captivated the hearts of the deer, snakes, lions, elephants, tigers, etc. The snake and the peacock, the lion and the elephant, shed their enmity and lived together happily. The rivers stopped in their course. The waves in the sea calmed, to hear the music of Ayanar. Even the celestials (Vidyadharas, Kinnaras and Devas,) came in their celestial cars to hear the music.

The Lord was immensely pleased with Ayanar's sincere devotion. The sweetness of the music of the flute and the effect of the Panchakshara Mantra both melted His heart. He appeared before Nayanar, with Mother Parvathi, blessed him and took him to Kailas.

Murthi Nayanar: To adore Lord Shiva with sandalwood paste, smearing it all over the Lingam is regarded as a great form of His worship. This kind of worship was done by Murthi Nayanar. He was born in Madura in Pandya kingdom. He was a Vaisya by caste. He was a great devotee of Lord Shiva. Daily he used to offer sandalwood paste to the Lord.

At that time, the city was invaded by a Karnataka king. In the battle the Pandya king was defeated. The Karnataka ruler became the Pandya king. He was a follower of Jainism. He wanted to exterminate Shaivism and to spread his religion. He began to persecute Shaivas. 'Murthi Nayanar also had to bear a lot of sufferings. But, he was undaunted. He continued his worship of the Lord, with sandalwood.

The king, with a view to convert Murthi Nayanar forcibly to Jainism, made it impossible for anyone in Madura to obtain sandalwood. This greatly upset the Nayanar. He prayed to the Lord: 'Oh Ocean of Mercy, this country is ruled by a tyrant and he is bent upon exterminating Your devotees. When will we be fortunate enough to get a king who will be devoted to You?' He knew that the people would follow the king, out of fear and in an effort to win his favour. He, therefore, wanted a Saivite king!

He searched throughout the day for a little sandalwood to offer his worship. He could not get any. With a broken heart, he went to the temple: and he had a wonderful idea. He began to rub his own elbow (in the place of sandalwood!). The hand was bleeding profusely. Lord Shiva was highly pleased with his devotion. A heavenly voice said: 'Oh noble soul, I am immensely pleased with your devotion. Kindly stop rubbing your elbow. All your grievances will be redressed. Kindly take up the reins of the kingdom. After ruling the country justly and wisely for a long time, you will come to My Abode.' Nayanar was amazed to hear this and to see that his elbow regained its original shape.

Murthi Nayanar did not aspire for kingship, but it was the Lord's will. That night the cruel king died. The next day, the Ministers sent the palace elephant to choose their king, in accordance with an ancient custom. The elephant proceeded towards the temple. Murthi Nayanar had come there for his worship. The elephant bowed to him and placed him on his back and returned to the palace.

The Ministers begged of Nayanar to become their king. Nayanar stipulated this condition: 'If I become king, I will not have any luxury bath, but will bathe only with the sacred ashes. My jewel will be only Rudraksham and my crown will only be matted locks. I shall strive to let the love of Lord Shiva be enthroned in the hearts of all.' The Ministers accepted these conditions with great joy and satisfaction.

Nayanar ruled the country justly and wisely for a long time, and eventually attained Shiva's Abode.

Muruga Nayanar: To do Archana with flowers mentioned in Shiva Agamas, to offer flower garlands to the Lord, and to repeat the Panchakshara Mantra is the ideal form of worship.

Panchakshara Mantra is considered a very great Mantra because it occurs in the centre of the Sri Rudradhyayi, which occurs in the middle of the middle Khanda of the Yajur Veda. He who does Japa of this Mantra is at once relieved from the disease of birth and death. Muruga Nayanar excelled in this worship. He was born in Tirupukalur, made famous by many other Nayanars also.

Daily he would wake up well before sunrise, take bath, wear the sacred ash on his forehead, do his Nitya Karmas, and go to the garden with a basket. Repeating the Panchakshara he would collect flowers, and as mentioned in the Shiva Agamas, make colourful garlands and offer to the Lord.

One day the great Jnana Sambandar came to his place. Muruga Nayanar invited him and worshipped him and won his favour. Sambandar took him as his dearest friend. Muruga Nayanar got the great good fortune of attending Sambandar's wedding when he, the bride and all others (including Muruga Nayanar) got merged in the Divine Effulgence of the Lord. Hence, the Kural says: 'Rarest of all rare things is to win the great to one's side by courting it.' Muruga Nayanar's devotion won for him Sambandar's friendship, and, through that, God-realisation.

Rudra Pasupathi Nayanar: There are seven Khandas in the Yajur Veda regarded as the Head of the Lord. Sri Rudram is in the centre of the middle Khanda of the Yajur Veda. Its recitation is a great purifier. It describes the Lord's wonderful manifestations. Standing in the river and reciting this sacred scripture is regarded as specially efficacious, and bestows Moksha on the devotee.

Rudra Pasupathi Nayanar was a great devotee of Lord Shiva and he resorted to this Sadhana. He used to recite Sri Rudram standing in water, neck deep and was, therefore, blessed by the Lord with Moksha.

Tiru Nalai Povar Nayanar: Nandanar was a Paraiah (untouchable) by caste. He was born in Adanoor in the Chola kingdom. He was an embodiment of humility and devotion. Lord Shiva was his sole refuge. He would often visit the holy places of pilgrimage, and supply leather drums and such other musical instruments for the temples. When he went near the temples, he always remained outside and worshipped the Lord mentally.

Once he had a desire to have Darshan of the Lord at Tirupunkur. He was also eager to do some service to the Lord there. He went to Tirupunkur and stood in front of the temple. He was grieved because Nandi which is always right in front of the Lord was hiding Him. Nandanar prayed to the Lord fervently. The Lord was highly pleased with his devotion and asked Nandi to move a little so that Nandanar might have His Darshan. Even today, at Tirupunkur, Nandi is leaning to one side! Nandanar had a delightful Darshan of the Lord. After digging a tank near the temple he returned to Adanoor.

The desire now arose in him that he should go to Chidambaram and have Darshan of Lord Nataraja. The love of the Lord had grown so intense that he would shed tears of love and tell his companions: 'I will surely go to Chidambaram tomorrow.' This expression earned for him the name Tiru Nalai Povar (one who would go tomorrow). One day he actually left his place and went to Chidambaram. He went round the village and, thinking of his low birth, did not want to enter it. He prayed: 'Oh Lord, I want to see your Cosmic Dance in Your Nritya Sabha. But, how can I? On account of my low birth they will not allow me to enter the temple.'

For days he went on praying like this. The Lord, pleased with his devotion, appeared in his dream and said: 'Oh noble soul, do not grieve. You will come to Me. Take a fire bath. Then come to My Kanaka Sabha along with the Brahmins.' Nandanar woke up and was highly pleased. At the same time, the Lord appeared before the Brahmins of Tillai, in their dream, and said: 'O Brahmins, My dearest devotee, Tiru Nalai Povar, has come to Tillai. Prepare a sacred fire. Nandanar will take bath in it and then come to Me.'

The next day, the Brahmins prepared the sacred fire. They went to Nandanar, prostrated before him and related their dream. Nandanar went round the fire, and with His Name on his lips and his mind fixed on the lotus feet of the Lord, he jumped into the fire. He emerged from the fire with a new holy body, with sacred ashes smeared all over, the holy thread and matted locks. He was then taken inside the temple. In the Kanaka Sabha, he worshipped the Lord. He went into a divine ecstasy and was completely absorbed in the dance of the Lord. A dazzling light was seen in the room, and Nandanar had disappeared. He had become one with Lord Nataraja.

Tiru Kurippu Thonda Nayanar: Thondamandalam was a prosperous land. Its capital was Kanchipuram. Here, Parvathi worshipped the Lord, according to the Agamas. The Lord here is called Ekambaranathar.

Tiruthondar was born here. He was a washerman. He was a staunch devotee of Lord Shiva. He served Shiva Bhaktas, understanding their need by watching for the signs on their face, and hence he had earned the name Tiru Kurippu Thonda Nayanar. His service consisted mainly of washing the clothes of Shiva Bhaktas. Lord Shiva wanted to bless this devotee: and, as usual, it had to be preceded by a severe test.

The Lord disguised Himself as a poor man, with Rudraksham on his neck and sacred ashes over the body, and appeared before Tiruthondar wearing a dirty rag. The very sight of the Shiva Bhakta put Nayanar in a trance. He worshipped him. Thondar then asked him: 'You have purified my house by your visit. How is it you are so emaciated? And, your rag needs washing. Kindly allow me to do tnis service for you.' The Shiva Bhakta agreed to let him do so on one condition: the rag should be washed, dried and returned to him before sunset, otherwise his emaciated body would perish in the cold.

When Thondar accepted the work, there was brilliant sunlight. He had washed the rag and immediately, it began to rain heavily. It was nearing sunset time. There was no hope of getting the rag dried. Thondar was greatly upset. Instead of serving the Shiva Bhakta, he was going to put him to great hardship. Thinking of this sin, Thondar, dashed his head, prayer fully, on the washing stone, and began to weep.

Lord Shiva appeared before him, held his hand and said: 'Oh noble soul, I am highly pleased with your sincere devotion. You will soon come to My Abode and live happily there forever.'

Tiru Kurippu Thondar fell at the Lord's Feet and sang His glory.

Chandeshvara Nayanar: Vichara Sarman (who was later known as Chandeshvara Nayanar) was born in Tiruchaijnanallur, which was famous for Vedic recitations, Tapas, and Shiva Bhakti. He was a prodigy. At the age of five he had learnt the Vedas, and all the Agamas, by himself, on account of previous Samskaras. After his Upanayanam (sacred thread ceremony), he learnt the Vedas, under a Guru: but the Gurus were wonder-struck at the intelligence of the disciple. He wanted to attain Final Emancipation in that birth.

One day Vichara Sarman, with his friends, was walking along a road. He noticed a cowherd severely beating a cow because it had slightly pushed him with its horn. Vichara Sarman could not endure this. He was greatly moved and spoke to the cowherd: 'Oh ignorant man? Do you not know that the cow is worshipful and divine. All the Devas dwell in her. She is indispensable for all religious activities. It is our sacred duty to tend and protect the cows. Whoever harms the cow is hurled into the Naraka. Whoever worships the cow wins a place in heaven or in Shiva's Abode. You have committed a great sin today. Hereafter, you need not tend to the cows: I will do that myself.' Vichara Sarman took the work upon himself from that day.

The cows grew healthier. They liked Vichara Sarman very much. The Brahmins who got more milk thereafter were able to fulfil their religious duties very well, and they were pleased with Vichara Sarman. There was so much of surplus milk now, that Vichara Sarman, who loved Lord Shiva and His worship, decided to perform Abhishekam for Lord Shiva. He used to sit under an Atti tree on the bank of the river, construct a temple from the mud, and also make a mud-image of Shiva Lingam and offer the milk, to bathe the Lingam. Then he would perform Archana with the Atti flowers. The cows were giving plenty of milk both to Vichara Sarman and to the Brahmins.

One day, when Vichara Sarman was doing this Puja, a villager happened to pass by, and he watched all this. He reported the matter to the owners of the cows, and complained that Vichara Sarman was pouring precious milk on mud and river sand. These Brahmins summoned Vichara Sarman's father, Echhadattan, and passed the complaint on to him in rather strong terms. The father was naturally shocked. He came home, but did not speak to Vichara Sarman. He wanted to find out for himself. So, the next day, he followed Vichara Sarman at a distance without his knowledge. When Vichara Sarman began pouring milk on the improvised Shiva Lingam, the father, without understanding the son's wonderful devotion, hit him with a stick.

The boy was so much absorbed in his Puja that he did not even feel pain. Then, the father, getting still more angry, knocked the milk-pot down. It was then that Vichara Sarman realised that his father was interfering with his worship and had committed an unpardonable offence against Lord Shiva (Shiva Aparadham). He at once took a stick and hit his father's leg: by the will of the Lord, the stick was transformed into an axe. Again, by His will, it killed the father. It was to test the depth of Vichara Sarman's devotion. He was so engrossed in the worship, that he did not mind what had happened and continued the worship.

Lord Shiva was immensely pleased with the intensity of Vichara Sarman's devotion and appeared before him, with Parvathi. Vichara Sarman prostrated before the Lord. The Lord embraced Vichara Sarman and fondled him. That very instant, Vichara Sarman attained the divine Form of Lord Shiva. The Lord removed a garland from His own neck and put it around Vichara Sarman's. He had attained Saroopya Mukthi (liberation, with the attainment of the form of the Lord). The Lord said: 'My child, you cut your own father's leg for My sake. Now, I am Your Father. You will soon attain Chandikesvarar's Abode.

You worshipped Me with food, clothes and flower garlands. In the same manner, you will also be worshipped.' The Lord disappeared. Vichara Sarman also went to the Abode of Chandesvarar. No sin attached to him, for having killed his own father, because of his supreme devotion to the Lord. His father, too, because he was killed by such a great devotee of the Lord, was purged of the sin of interfering with His worship, and reached the Abode of Lord Shiva.

Tiru-Navukkarasar Nayanar: Appar or Tiru Navukkarasar flourished in the 7th century A. D. He is one of the four Shaiva Samaya Acharyas (Saivite spiritual teachers). He was born in Tiruvamoor in Tirumunaipadi Nadu. Pukalanar was his father; Mathiniyar, his mother. Mathiniyar gave birth to a daughter whom they named Tilakavathi. After some years, Mathiniyar had a son whom they called Marulneekiar, the dispeller of darkness or ignorance. Early in life he mastered all the Shashtras.

When Tilakavatiar reached her twelfth year, she was betrothed to Kalaipahayar, a military commander in the Pallava army. Before the wedding, however, he was sent by the king to fight another, and he died in battle.

Pukalanar fell seriously ill and died. Mathiniyar committed Sati (died on the husband's funeral pyre). As the children were recovering from this shock, the news of Kalipahayar's death reached Tilakavathiar: and as she, since her betrothal, had regarded him as her husband, she decided to commit Sati, too.

But, Marulneekiar pleaded with her reminding her that now she was his mother, and also threatened to die if she would not change her mind and live. Tilakavathiar changed her mind for the young brother's sake. Even though she was young, she led the life of an ascetic. She was highly devoted to Lord Shiva. Her glorious ascetic life has been sung by Sekizar, the author of Tirumurai. She was mother to Appar.

Marulneekiar, even while young, had realised the unreality of the world. He engaged himself in all kinds of charitable works. He was eager to find out the best religion and to follow it. He had heard much about Jainism and its wonderful practice of Ahimsa. He believed that Jainism would give him emancipation and so became a convert. He even went to Pataliputra (in South Arcot district) and joined the Jain school. He attained mastery over all their scriptures.

Tilakavathiar was heart-broken over this change in her brother. She abandoned her native place and settled in Tiruvadigai Virattanam, in a Mutt she built there. She prayed fervently to Lord Virattaneswarar to save her brother and shower His grace upon him. The Lord appeared in her dream one day and said: 'My child, your brother has already done severe Tapas in order to attain Me. I will surely turn his mind, by making him suffer from severe colic, and then take him to My fold.'

Marulneekiar fell a victim to severe colic. He could not bear the pain. The Jains tried their best but could not relieve the pain. He felt intuitively that it was an eye-opening experience. He lost faith in Jainism. He thought of his sister. He threw away the Jain garb and without informing anybody, returned to his sister. He fell at her feet and prayed to her to protect him. She understood it was His Lila, and said that by the grace of Lord Shiva, he would be all right. She smeared the holy ashes on his forehead and repeated the Panchakshara Mantra. His ignorance vanished.

She took him to the temple of Virattanesvarar. He worshipped the Lord and sang a hymn: 'Oh Lord, I have insulted You and Your religion. I have committed many evil acts. Once on the bank of Godavari, I argued with the saints and established the superiority of Jainism. For all this evil, Lord Yama himself has come to me in the form of this excruciating pain. Oh Lord, You are my sole prop and refuge. Save me. I will ever keep Your Lotus Feet in my heart.' When he concluded the song, the pain disappeared. A celestial voice said: 'From now on you will be known as Tiru Navukkarasar, 'Lord of Speech'. Your glory will spread everywhere.' Thus Lord Shiva's grace restored his faith in Shaivism. Tilakavathiar was immensely happy, too. Tirunavukkarasar became a staunch devotee of Lord Shiva and lived in Him repeating the Panchakshara Mantra.

The Jains at Pataliputra were afraid that, if the king came to know that because of their inability to treat Tirunavukkarasar (to whom they had given the name Dharmasenar) he had left them and gone back to Shaivism, he would take them to task. So they concocted a new story and showed Tirunavukkarasar as a traitor against the king and the royal religion, *viz.*, Jainism. The king ordered his ministers to produce Tirunavukkarasar before him.

They went to Tiruvathikai, with an army. When Tirunavukkarasar heard of the charge against him, he said: 'Oh Ministers! I am no longer your king's subject. I am the subject of Lord Shiva, the protector of all beings, the destroyer of all sins, the Lord of all gods, the bestower of immortality and eternal bliss. Disobedience to the king may amount to treason in the case of other people, not mine, because I am under His protection. Fear cannot approach me, because I am under the protection of one who once kicked Lord Yama, to save His (Shiva's) Bhakta.' The Ministers recognised his greatness: but were afraid to return without him. They, therefore, begged of him to come with them out of his supreme compassion, and to establish the glory of Shaivism. Tirunavukkarasar went with them.

The very sight of Tirunavukkarasar enraged the king who asked the Jain heads to decide upon the proper punishment to be meted out to him. They suggested that he be thrown into a burning lime kiln. Accordingly, he was shut up in a kiln for seven days. He remained there, fixing his mind on the Lord and repeating His Name. By His grace, the heat of the kiln was transformed into a cool breeze. At the end of seven days, the Jains, to their surprise, found Tirunavukkarasar alive, and absorbed in deep meditation.

They attributed this to the power acquired by him when he was a Jain: and advised the king to poison him. Again, by the grace of Lord Shiva the poison was transformed into nectar.

Again, the Jains attributed this to the efficacy of the Jain Mantras which Tirunavukkarasar had learnt from them, and advised the king to have him trampled by the elephant. Tirunavukkarasar, boldly facing the elephant sang a hymn in praise of the Lord. Tirunavukkarasar's loving look transformed the elephant's nature and it went round him and prostrated to him. The Mahouts goaded it: but it got wild and attacked the Mahouts and the Jains and killed some of them. Those who escaped ran to the king and fell at his feet. This was a great disgrace.

The king was greatly worried. The Jains finally advised the king to have Tirunavukkarasar tied to a stone and thrown into the sea. In accordance with the king's orders, this was done. Tirunavukkarasar fixed his mind on Lord Shiva and was continually repeating the Panchakshara. He sang hymns in praise of the Panchakshara. As soon as he finished the song the stone began to float. Tirunavukkarasar sat on the stone and was happily borne on the waves and safely taken to the shores of Tiruppapuliyur. Thus did the Lord save His Bhakta.

At Tiruppapuliyur, there was a huge congregation of Shiva Bhaktas to welcome him. He worshipped the Lord and sang hymns in praise of the Lord. 'Oh Lord, the unseen Protector: You are my guide and saviour. You are my father, mother, sister and everything. Oh Lord of Mercy, You saved me from all dangers. Due to Your grace and love alone I am alive.' Addressing the mind, he says: 'Oh mind, when you have totally surrendered yourself to Him, why do you fear any danger? None can harm you. Fear not.' Again, turning to the Lord, he says: 'Oh Lord of Mercy, I want no more birth. If I take birth at all, owing to past Karma, let me remember Your Name always. Even if I take birth as a worm, let me not forget Your Name.

Let me find delight in uttering Your Name.' Then Tirunavakkarasar returned to Tiruvathikai, after visiting many holy places on the way. When Pallava king who persecuted him came to know this, and when he recalled the many miracles he had witnessed, he was convinced of the superiority of Shaivism. He went to Tiruvathikai, fell at Tirunavukkarasar's feet, and begged his pardon. Tirunavukkarasar embraced the king with all love and affection, and the king embraced Shaivism and built the magnificent temple of Shiva called Gunabharaveechuram at Tiruvathikai.

Tirunavukkarasar then spent his days in worshipping and serving the Lord in various ways, to set an example even to saints that they should not relax their eternal spiritual vigilance, lest they should fall a prey to Maya and to exhort them to lead the ideal life of a humble devotee for the guidance of others. He also visited many sacred shrines, thus emphasising the glory of pilgrimage.

The thought that he had lived for some time with the Jains, eating their food and mixing with them made Tirunavukkarasar feel that his body was still impure, for the worship of Lord Shiva. He prayed to Lord Shiva: 'I do not want to live any more in this impure body. Let me have the stamp of Your Trident and Nandi on my body, and then I shall regard it as fit for Your worship.' At once a divine servant of Lord Shiva approached Tirunavukkarasar and put the stamp of the Trident and Nandi (one of the Saivite rituals of initiation, according to the Agamas) on Tirunavakkarasar's shoulders. He experienced supreme bliss immediately.

Tirunavukkarasar then went to Chidambaram. The very sight of the temple tower sent him into a trance. He sang thrilling hymns here and prayed: 'Let me serve You. Your Bhaktas know no want nor fear. Even Lord Yama cannot dare to approach them. I have come to Your Abode where no sin can approach. I have surrendered myself to You. I have enshrined Your Lotus Feet in my heart.'

At Chidambaram, he heard of the glory of Tiru Jnanasambandar and how he was blessed by Parvathy. He was eager to meet the great saint who was at Shiyali. As Tirunavukkarasar was proceeding towards Shiyali, Sambandar also was happy and was eagerly looking forward to this meeting. When Tirunavukkarasar reached the outskirts of Shiyali, Sambandar went forward to receive him. They fell at each other's feet, and set an example in saintly conduct. They went to the temple and worshipped the Lord. At Sambandar's request,

Tirunavukkarasar sang a hymn here: 'When the whole world was submerged during Pralaya, Lord Thoniappar was seated with His Consort in the boat of Pranava (OM) surrounded on all sides by Devas in the form of birds.' Even now the temple of Lord Thoniappar is in the form of a boat. They who cling to His feet are protected.

Along with Sambandar Tirunavukkarasar visited the temple at Tirukkolaka. He then took leave of Sambandar and left the latter's Ashram and after visiting many shrines, reached Tiruvavaduthurai. He sang a hymn here expressing intense Vairagya: 'Oh Lord, I am caught in the wheel of births and deaths. I am tired of this. Show me a way to get out of it.'

Then he came to Sattimutham. He prayed to the Lord: 'Oh Lord, place Thy Lotus Feet on my head before I leave this body.' The Lord said: 'Come to Tirunallur. I will fulfil your wish.' Tirunavukkarasar accordingly went to Tirunallur and worshipped the Lord Who placed His Feet on the head of Appar. He sang: 'Oh Lord, Thy Holy Feet are adored by men on earth and the gods in heaven. Even great Tapaswins cannot touch Your Feet, if their devotion to You is tainted. But They are easily accessible to the real devotees who serve with faith and devotion. The sacredness of Your divine Feet cannot be understood by ordinary men. They are the bestowers of all prosperity and immortal bliss.'

After visiting some more places of pilgrimage, Tirunavukkarasar went to Tingalur, where he formed a miracle and brought saint Appudi Adigal's son back to life—we shall read this in Appudi Adigal's life.

At Tiruvarur, he was given a rousing welcome by the Shiva Bhaktas. He had the Lord's Darshan there. His heart was overflowing with love. He shed tears of love. He danced in joy. He was immersed in divine bliss. He sang hymns expressing regret for his mistake in joining Jainism. He felt that he would have been blessed with the Lord's Darshan earlier had he remained a Saivite and carried on His worship, with devotion. He also sang a hymn praising the sincere devotion and greatness of Nami Nandi Adigal who lighted the lamp with water in this sacred shrine when he could not get oil anywhere.

Then Tirunavukkarasar went to Tirupukalur, after visiting other shrines on the way. Tiru Jnana Sambandar was there, too, and the two saints met each other for the second time. At Appar's instance, Sambandar went to Tiruvarur and had the Lord's Darshan there. Both the saints remained at Tirupukalur for some time: and many other saints took this golden opportunity of having the Darshan of these two great Acharyas (spiritual preceptors) together. Tiru Neelakantha Nayanar, Siruthondar, Muruga Nayanar and many others came to Tirupukalur, and the place was converted into a divine realm during the stay of the two Acharyas.

The two saints then moved on to Tiruveezhimalai. Famine raged there at that time. Appar (another name for Tirunavukkarasar) and Sambandar were greatly moved by the suffering of the people and prayed to the Lord for relief. The Lord promised to give them a golden coin each every day, with which they could feed the people. Lord Shiva placed two coins, one at the western entrance and the other at the eastern entrance to the temple: the Nayanars collected the coins and relieved the suffering of the people. The famine soon came to an end. Both the saints praised the glory of the Lord and left.

Then they came to Tirumaraikadu or the present Vedranyam. Once upon a time, the Vedas themselves used to worship the Lord here. When people neglected the study of the Vedas, this was discontinued. From that time, the door by which the Vedas used to enter, remained closed. There was another door by which people would go in and worship. Appar and Sambandar heard of this when they came there. Sambandar desired to enter through the door which remained closed. He requested Appar to sing a song.

The doors opened by themselves, by the Will of God. They went in and worshipped the Lord. When they returned to that entrance, Appar requested Sambandar to sing a song, so that the door could close again. Sambandar sang and the door closed again. At night one day, Lord Shiva appeared and commanded Appar to come to Tiruvaimoore. The Lord appeared before him and walked in front of him. Appar followed Him, but could not approach Him. Suddenly the Lord entered the local temple and disappeared. Appar went inside the temple but could not find the Lord. In the meantime, Sambandar, learning of Appar's departure, followed him and came to the temple. Appar entreated the Lord to bless Sambandar with His Darshan. The Lord fulfilled Appar's wish. Then, they went to Tiruvaimoore and from there returned to Tirumaraikadu.

Mangayarkarasiar and Kulachirai Nayanar, the queen and the minister of the Pandyan king, sent messengers from Madurai to Sambandar, reporting the evil influences of the Jains and urging for his immediate presence in Madurai. Sambandar wanted to go there immediately. Appar, in the meantime, told him of all that happened to him and tried to stop him from going. But, so great was Sambandar's eagerness to serve the Lord, that he went.

Tirunavukkarasar then visited Tiruvavaduthurai, and came to Pazhaiyarai. Here he came to the Vadathalai temple and worshipped the Lord from outside. He came to know that the Jains had converted this into a Jain temple, and that they had removed the Shiva Lingam to an unknown place. He prayed to the Lord: 'Oh Lord, I will not proceed an inch from here till I have the Darshan of Your image which has been removed by the Jains to an unknown place.' The Lord appeared before the king in his dream and said: 'Oh king, My Bhakta Tirunavukkarasar is fasting, to have My Darshan. Go at once and drive out the Jains, so that Tirunavukkarasar could enter the temple and have My Darshan.' He also told the king where the Lingam lay hidden.

The king woke up and immediately summoned his ministers. He went to the temple, drove away the Jains and fell at the feet of Tirunavukkarasar. The temple was immediately reconverted into a Shiva temple and the Lingam installed once again. Tirunavukkarasar worshipped the Lord and was happy.

During another pilgrimage, Tirunavukkarasar felt the pangs of hunger as he was approaching Tirupainjeeli. Lord Shiva wanted to appease his hunger and thirst. He created a tank and a garden on the way, so that Tirunavukkarasar could quench his thirst and rest in the garden. The Lord Himself waited there in the guise of a Brahmin, with food in hand and gave it to Tirunavukkarasar as soon as he arrived there. Appar took the food, drank the water and was resting when the Brahmin enquired where he was going.

He said that he was going to Tirupainjeeli. They both started to walk. When they were near the place, the Brahmin suddenly disappeared, and Appar understood that it was none other than the Lord Himself. He wept bitterly for not recognising Him earlier and rolled on the ground on account of His separation.

After visiting Tiruvannamalai, Kancheepuram, and Kalahasthi, where he sang the glories of Kannappar, Appar felt a desire to go to Kailash. He went to Banaras, and worshipped Lord Viswanath. He turned northwards and crossed many thick forests infested by wild animals. By his mere look, the wild animals became tame! He walked night and day. His feet were sore. Then he crawled with his hands. His elbows began to bleed now. Then he used his chest and crawled on. His chest also began to bleed and the ribs began to break. Still, Appar, undaunted, continued his journey to Kailash, rolling on the ground. Appar wanted to go to Kailash: but the Lord wanted that he should live in the world for many more years singing His glories.

The Lord created a tank nearby and appeared before Appar in the form of a saint, with matted locks, Rudraksham and holy ashes. The saint found out from Appar that he was proceeding to Kailash, and said: 'Oh friend, the Lord of Kailash cannot be seen by human beings. So, turn back.' Appar said: 'So long as I have this body, I will not turn back, without going to Kailash,' and turned towards the saint again after bowing to him: but the saint was not there. Appar understood it was the Lord Himself. The Lord afterwards kept Appar invisible company, giving him encouraging words now and then. Appar mentally prayed to the Lord: 'Oh Lord, give me a fresh body, so that I may continue the journey.'

The Lord directed him to take bath in a near-by tank and said: 'You will see Me and My Abode, Kailash in Tiruvayar.'

Appar, repeating the Panchakshara, took a dip in the tank. He came out of the tank, and found himself in the tank at Tiruvayar, hundreds of miles away, to the south! He came outside and saw everywhere Shiva and Shakti. He entered the temple and saw Mount Kailash there. He saw Lord Shiva seated with Mother Parvathy, surrounded by gods, and celestial servants praising His glory. He went into a trance and sang His glories and danced.

Then, Appar had a Mutt built for him at Tirupoonthurai and remained there. Sambandar had, in the meantime, defeated the Jains at Madurai and was coming to Tirupoondurai. Appar went forward to receive him. Without Sambandar's knowledge, Appar quietly joined those who were carrying his palanquin. On reaching Tirupoonthurai, Sambandar cried out: 'Where is Appar?' and Appar, from below the palanquin coolly announced himself. At once Sambandar jumped out of the palanquin and fell at the feet of Appar who had, by his example, demonstrated the humility of a true saint. They embraced each other and shed tears of love. (Incidentally, it is interesting to note that Appar was advanced in age and Sambandar was only seven years old at the time.)

Then, Tirunavukkarasar wanted to see the state of Shaivism in the Pandyan kingdom, for himself and left for Madurai. The king, Ninra Sheer Nedumara Nayanar, the queen Mangayarkarasiar, and the minister Kulachirai Nayanar welcomed him with devotion. Appar remained there for some days, worshipping the Lord. Then he went to Rameswaram and other sacred places before returning to Tirupukalur.

Lord Shiva wanted to test him here. When Appar was doing his services in the temple, the Lord made the entire floor appear as though it was strewn with gold and diamonds. To Appar, gold and diamonds were worthless 'straw'. He collected all of them and threw them in a near-by tank. Again, the Lord made celestial damsels appear before him and tempt him with their charms. Appar remained undisturbed. His entire heart and soul was centred on the Lotus Feet of the Lord.

Appar spent the rest of his life there and at the age of 81 merged himself in Lord Shiva.

Kulacchirai Nayanar: In devotion to Shiva Bhaktas, Kulacchirai Nayanar excelled. He was born in Manamerkudi, in the Pandyan kingdom. This place was frequently graced by Shiva Bhaktas. Kulacchirai Nayanar was the leader as well as the supporter of the people. To him adoration of Shiva Bhaktas was equal to adoration of Lord Shiva Himself. He saw no difference between Shiva Bhaktas and Shiva.

He was the Prime Minister of the Pandyan king. Yet, he regarded himself as the slave of Shiva Bhaktas. He was the richest man in the place: yet, to him wealth was only dust. Nothing belonged to him: it was the property of Shiva Bhaktas. Even Sambandar extolled his virtuous qualities in a song.

Nayanar was an able soldier and administrator. Yet, his mind was ever absorbed in the Lord. He helped the queen in stemming the tide of the evil influence of Jainism. Nayanar invited Sambandar to Madurai to fight this evil influence. The Jains set fire to Sambandar's camp. Sambandar sang a song. The fire was extinguished. The Pandyan king had high fever, which the Jains could not cure, but which was cured by the sacred ash which Sambandar applied on him. Sambandar argued with the Jains and defeated them. Kulacchirai sent the defeated Jains to the gallows. He served the Shiva Bhaktas and finally attained Shiva's Abode.

Perumizhalai Kurumba Nayanar: Guru is God. The same Lord Who is never separate from us, Who is our sustainer and support appears to us as the visible form of the Guru. He who adores the Guru with faith and devotion will attain all Siddhis (psychic powers) and eternal bliss. Perumizhalai Kurumba Nayanar excelled in Guru Bhakti. He was an ardent devotee of Lord Shiva and Shiva Bhaktas, too. He heard of Sundaramurthi Nayanar's greatness and mentally accepted him as his Guru. To him, Sundarar was the sole refuge. He adored the Guru in thought, word and deed. By the Guru's grace, he attained all the Siddhis. He was immersed in Shiva Bhakti and Guru Bhakti.

In the meantime, Sundarar came to Tiruvanchaikalam from where he was taken to the Lord's Abode. Kurumba Nayanar, through his Yogic powers, came to know that this would happen.

He did not like to remain in this world after the Guru: and, therefore, through the method of Shiva Yoga, Nayanar cast off his mortal coil and reached the Abode of Shiva, a day before Sundarar's departure.

Karaikal Ammaiyar: Punithavathiar as Karaikal Ammaiyar was called, was born in a Vaisya family. Her father was Danadathan. He was a wealthy merchant. He was very virtuous, too. He and his dutiful wife prayed to the Lord for a child, and the child the Lord blessed them with they called Punithavathy. From her childhood, Punithavathy had an intense love for Lord Shiva and His Bhaktas. She was married to Paramadattan, a wealthy Vaisya. Both of them were leading an ideal householder's life.

One day Paramadattan sent two mangoes to his house. Punithavathy kept them safely so that she could serve her husband with them at meal-time. In the meantime, a Shiva Yogi appeared before her. He was hungry and completely exhausted. Punithavathy worshipped him and offered him Bhiksha. She had nothing to give him, except the mangoes. She gave one to the guest. At midday Paramadattan came to the house. The wife served him with one mango. He liked it, and asked for the other. She was upset. She appealed to the Lord for help. When she finished her prayer, mysteriously a mango fell on the palm of her hand. She gave it her husband. He tasted it. It was exceptionally sweet. He asked her to tell him from where she got it, as he was quite sure it was not the mango he had sent.

Punithavathy told him the whole truth. Paramadattan, however, would not believe this and challenged her to produce another. She prayed again to the Lord. Another mango appeared on her palm. She gave it to him. But, at once it disappeared from his hand. He was astounded. He understood the greatness of his wife. He felt that it was a great sin to live with her as her husband. On the pretext of going to a foreign country for trade, he sailed with a ship load of goods. On return, he established himself in a big city in the Pandyan kingdom. He married a Vaisya girl and lived happily. He had a daughter by her and he named her Punithavathy, after his first wife.

Punithavathiyar's relatives came to know of her husband's whereabouts and took her also there in a palanquin. When Paramadattan heard that Punithavathy was coming to him, he, with his second wife and child, went forward, and fell at Punithavathy's feet. When the people demanded an explanation, he revealed that he regarded her, not as his wife, but as a Goddess. Punithavathy understood his mental condition, and prayed to the Lord: 'In that case, Oh Lord, deprive me of the present physical charm and let me have a demonaical form.' Her prayer was immediately granted and her charming body was transformed into a skeleton.

Then she went on a pilgrimage to the holy Kailash. Feeling that it would be a great sin to place her foot on those sacred grounds, she made the last part of the journey on her head. Mother Parvathy was surprised to see Punithavathy's strange form and her wonderful devotion. Lord Shiva told her of Ammaiyar's greatness. When she went near Him, Lord Shiva welcomed her with extreme love and granted a boon to her. She fell at His Feet, and prayed: 'Oh Lord of Mercy, give me sincere, pure, unalloyed, eternal and overflowing devotion unto You. I want no more birth.

If, however, I have to take birth here, grant me that I should never forget You. Whenever You dance, I must be at Your feet singing Your praise. This is my only wish.' Lord Shiva granted the boon and asked her to proceed to Tiruvalangadu to witness His dance. She went to that place and spent her life singing the praise of Lord Shiva.

Appuddi Nayanar: In devotion to the Guru, Appudi Nayanar excelled. He was an ardent Shiva Bhakta. He was leading the ideal householder's life. He belonged to a Brahmin family in Tingalur in the Chola kingdom.

Appudi had heard of the glories of Tirunavukkarasar or Appar. He had heard of how God's grace made the stone float and how Appar rode on it and floated on the sea and went to a place of safety. Even though he had not seen Appar, he had taken him as his Guru, and literally worshipped Appar. He knew that Lord Shiva Himself, out of compassion for the spiritual aspirants, appeared as the Guru. He meditated on the lotus feet of the Guru. He had named all his children 'Tirunavukkarasu': and all the household articles had also been named after the Guru. He had erected a number of water-sheds, for the service of pilgrims, and had named all of them after the Guru. Thus had he ensured that he would constantly remember the Guru, and experience his grace.

Appar himself passed through Tingalur one day. He went into one of the water-sheds. He was surprised to see his own name everywhere. He found out from some other pilgrims that the shed had been erected by Appudi and went to meet him. Appudi received the Shiva Bhakta (though he did not know who it was) with great devotion. Appar said: 'Oh noble soul, I have heard a lot about your greatness and glory. I wanted to pay my respects personally to you. Please tell me, why have you named the water-shed after somebody, and not yourself.' Appudi was upset at this casual reference to the blessed name of his Guru. He said: 'Oh friend, though you appear to be a Shiva Bhakta, you do not seem to know Tirunavukkarasu Swamigal, who through the grace of the Lord withstood successfully all the persecutions of the Pallava king and re-established Shaivism. Have you not heard how the king tied him to a stone and threw him into the sea, and how he floated back to the shore? Who are you?'

Appar was very much moved by Appudi's devotion and replied: 'I am that humble soul who fell a victim to severe colic and then took shelter under the Lord's Feet. I am that humble soul who, due to the grace of Lord Shiva, got cured of that disease and returned to Shaivism.' Look at the difference between the two descriptions! Appudi remembers the glory of Appar: whereas Appar chooses to recall his own failing (to preserve his humility) and the Lord's supreme saving grace.

As soon as he heard this, Appudi understood that the Shiva Bhakta was none other than Appar and was overjoyed. He worshipped Appar, along with his wife and prayed to Appar to accept his Bhiksha (food). While their son had gone to the garden to bring a banana leaf, for Appar to use as his plate, the boy was bitten by a cobra. The son of a Nayanar: he was also a great devotee of the Lord! He ran to the mother eager to fulfil his duty. He handed the leaf to his mother and immediately fell down dead. Appudi did not want to let this disturb his worship of Appar Swamigal: and, therefore, hid the corpse. He invited Appar to have his meal. Appar sat down and blessed Appudi and his wife with Bhasma, and then called for their son. Appudi tactfully replied: 'He is not in a position to come.' Appar sensed that there was something wrong and asked Appudi to tell him the truth.

Appudi informed him what had happened. Immediately Appar got up and asked Appudi to lay the corpse in front of the temple: and he himself sang a song. A miracle took place. The boy got up, as if from sleep. All were happy, except the parents of the boy. They regretted that this incident had caused some delay in Appar having his meals! Such is the nature of true devotion. Appar immediately took his meals and blessed the family.

Appar lived in Appudi Adigal's house for some time. Appudi gained the grace of the Lord, by his wonderful devotion to his Guru, Appar Swamigal.

Tiruneelanakka Nayanar: Tirusattamangai was an important city in the Chola kingdom. It was a place full of spiritual vibrations and Shiva Bhakti. The Brahmins were devoted to the study and recitation of the Vedas and worship of the Lord. And, the women were devoted to their lord (the husband) and served them as they would serve God Himself. It is situated seven miles east of Nannilam. There is a temple in this place called Ayavanthi. The Lord presiding over this is Ayavanthi-Nathar. His Consort is Malarkanni Ammai.

In this city there lived the glorious Brahmin, Tiruneelanakka Nayanar. He was well versed in the Agamas and was regular in his ritualistic worship of the Lord. On a Tiruvathirai day he was devoutly worshipping the Lord in the temple. His wife was also with him. A spider fell upon the Shiva Lingam, when the worship was in progress. The saint's wife, without a moment's hesitation, blew it away, and spat on the spot where it had fallen, on the Shiva Lingam—this is what they do when a spider falls on the body of a child or other human being. But, the husband was enraged at the wife's sacrilegious action: she had spoilt the worship and polluted the temple by spitting on the Lingam. Without a second thought, he abandoned her and returned home.

The lady appealed to Lord Ayavanthinathar for his mercy. He appeared in the saint's dream that night and showed him His body—all the parts of His body except that on which his wife had spat, had been affected by the spider poison. He realised that Bhakti was superior to ritualistic worship. He recalled the glimpse of the Lord he had, in his dream and rejoiced, rolled on the ground, wept out of sheer joy and danced. The next morning he went to the temple and worshipped the Lord, and returned home with his wife, the noble devotee to whom the Shiva Lingam was not a stone, but a Living Presence.

Once Tiru Jnana Sambandar visited his place with Tiruneelakanta Perumbanar and Virali. Tiruneelanakka Nayanar was very eager to meet the great saint Sambandar. He welcomed the saint with due honours. That night, Sambandar asked Tiruneelanakkar to give some accommodation to the other two who were with him.

They were not of the 'high' (Brahmin) caste! Tiruneelanakka Nayanar hesitated to let them sleep inside the house. He asked them to sleep near the sacrificial pit. As soon as they went near the pit, the Nayanar was astounded to see that the sacrificial fire began to burn of its own accord. He understood that, through the fire of their devotion they had attained to a stage which was far higher than what mere ritual could lead to. The mist of caste distinction also vanished from the eyes of Nayanar. Next morning, Sambandar went to the temple and sang a song in which he glorified Tiruneelanakka Nayanar also. When Sambandar wanted to leave the place, Tiruneelanakkar also wanted to accompany him: but Sambandar instructed him to stay there itself and serve the Shiva Bhaktas. He obeyed.

He was, however, longing to be always at the feet of Sambandar. Soon his wish was fulfilled. He heard of Sambandar's marriage, and went to Nallur Perumanam to witness it. When Sambandar got merged in the Light of Lord Shiva, Tiruneelanakka Nayanar also got merged in it.

Tiru Jnana Sambandar: In sacred Sirkali (which, according to a legend was the Noah's Arc during a cosmic dissolution) there lived a pious Brahmin by name Shivapada Hridayar with his virtuous wife Bhagavathiar. Both of them were ardent devotees of Lord Shiva. They refused to embrace Jainism and give up Shaivism, even though the forces of Jainism were powerful and devastating. Shivapada Hridayar prayed to the Lord for the boon of a worthy son to him who would reestablish the glory of Shaivism.

The Lord granted this boon, and Bhagavathiar soon brought into this world a radiant male child. They brought up this child with great love and devotion, knowing fully well that it was a purposeful gift from the Lord. The child, too, would weep for his separation from his divine parents Lord Shiva and Parvathy, though ordinary people mistook it for a baby's crying habit.

One day Shivapada Hridayar and his wife took the child with them to the temple tank in which they wanted to bathe. The child had insisted on being taken with them. They left the child on the bank and went in to bathe. The child looked at the tower of the temple and began to cry for his parents. This outwardly appears to be a mere childish action, but the Lord knew its inner meaning.

Lord Thoniappar wanted to bless the child. So, He appeared with Mother Parvathy and asked Her to feed the child with the milk of divine wisdom. To obtain His grace and divine knowledge, the grace of the Mother is necessary, Mother Parvathy fondled with the child and suckled him with the Milk of Wisdom. From that moment he was known as Aludaiya Pillayar or one who enjoys the protection of the Lord: and also as Tiru Jnana Sambandar as he attained divine wisdom through the grace of Lord Shiva and Parvathy. From the moment he drank the Milk of Wisdom, he began to sing soul-stirring songs in praise of Lord Shiva. The collection of these songs is called Thevaram.

After finishing their bath, the parents came to the child, and found a golden cup in his hands (the cup in which Parvathy gave him the milk) and milk overflowing from his mouth. Shivapada Hridayar thought that somebody had given milk to the child: he did not like that his child should accept milk from all sorts of people. So, he brandished a cane before the child and asked him who gave the milk.

The child, shedding profuse tears, pointed to the Lord Who appeared in the sky along with Mother Parvathy. He also sang a song in praise of the Lord. Shivapada Hridayar could not see the Lord, but guessed from the child's behaviour that he must have had a vision of the Lord. He followed the child into the temple, as he went towards it. Many devotees had also come to the temple. They had come to know of what had happened to Pillaiyar and glorified him. The parents were very happy. They took the child on their shoulders and went round the town in a procession. The people had decorated the town nicely and received Sambandar with great devotion.

The next day Pillaiyar went to Tirukkolakka and sang a song, clapping his hands to keep time. Lord Shiva, pleased with this, presented him with a pair of golden cymbals. Sambandar began to sing, with the help of the golden cymbals. Even Narada and the celestials were charmed by this. Sambandar then went on pilgrimages. Once Tiru Neelakantha Yazhpanar, an ardent devotee of the Lord and an expert musician on the Yazh (Veena) met Sambandar. They all went to the temple. Sambandar requested Yazhpanar to play the Yazh. The music melted the heart of Sambandar. Yazhpanar wanted to be always with Sambandar, to play on his instrument the songs that Sambandar sang in praise of the Lord. Sambandar granted this wish.

Sambandar went on a pilgrimage to Chidambaran. The very sight of the Lord entranced him. He had heard about the greatness of the Brahmins of Tillai (Chidambaram). To him, they actually appeared as Shiva Ganas (celestial servants of Lord Shiva). He showed this to Yazhpanar and they were thrilled. The Brahmins fell at his feet. Before they did so, Sambandar had fallen at their feet!

After visiting the birth-place of Yazhpanar, Sambandar wanted to go to Tiru Arathurai. He would sometimes walk and at other times sit on his father's shoulders. In this manner they approached Maranpadi. They were all tired due to the heat of the sun and the arduousness of the journey. They rested at Maranpadi for the night.

The Lord wanted to alleviate His child's suffering by presenting him with a palanquin. He appeared in the dream of the Brahmins of Tiru Arathurai and told them that they would find a pearl palanquin and a pearl umbrella, and asked them to take them to Sambandar who was then proceeding towards Tiru Arathurai. At the same time, the Lord appeared in Sambandar's dream and informed him of the gift! The next morning, the Bhaktas handed over to Sambandar the Lord's gifts to him. Sambandar worshipped the gifts and ascended the palanquin.

Sambandar returned to Sirkali, after visiting a number of shrines on the way, and singing Padigams in praise of the Lord everywhere. His parents performed the sacred thread ceremony. The Brahmins then began to teach him the Vedas. But, even before hearing the Vedas from the teacher, Sambandar could recite them, on account of previous Samskaras and divine grace. Then Sambandar taught them the essence of the Panchakshara and also sang a Padigam. It was at this time that Tirunavukkarasar also met Sambandar.

During the course of his pilgrimage, Sambandar came to Tiru Pachilasramam. The daughter of the Mazhava King there, who was a great devotee of Lord Shiva, was suffering from an incurable disease. The king had, in despair, taken her to the temple and placed her in front of the Lord. At the same time, Sambandar had come into the temple. He saw the pitiable condition of the girl, who was lying unconscious. He sang a Padigam praying for His grace upon the girl. She at once got up to the surprise of all. All were amazed at this miracle.

At Senkunrur, during his pilgrimage, Sambandar found that the cold was very severe and that many people suffered on account of it. They entreated him to alleviate their sufferings. Sambandar sang a song, and immediately, they were relieved of their suffering. After some more pilgrimages, Sambandar came to Tiruvavaduthurai. His father wanted to perform a big Yajna. He wanted a lot of money for that. Sambandar went to the temple and sang a song. At once a Shiva Gana appeared before him, handed him a purse containing one thousand gold coins and said. 'This purse has been given to you by Lord Shiva.' Sambandar glorified the Lord's grace, handed over the purse to his father (who went away to Sirkali) with the assurance that it would give inexhaustible wealth.

At Dharmapuram, which was the home of the Yazhpanar's mother, the people glorified Yazhpanar for his proficiency in music. Yazhpanar felt that it was due to Sambandar's grace that he was allowed to accompany Sambandar and that he could really not reproduce on the Yazh the divine melody of the saint's Padigams. To prove this Sambandar sang a song in praise of Lord Ganesha which Yazhpanar was unable to play on his instrument. He tried to break the instrument in desperation. But, Sambandar prevented him from doing so, and asked him to be content with what he could achieve with it, assuring him that was a lot.

Sambandar went to Sattamangai where he was received by Tiruneelakanta Nayanar with great love and devotion. Sambandar sang a Padigam in which he glorified the Nayanar. At Tiru Keizhvelur, similarly, he met Siruthondar and glorified him in a Padigam. Such is the nature of the truly great ones: they adore even devotees of the Lord as the Lord Himself and sing their glories, not regarding that as worship or adoration of a human being, but of manifest divinity.

During his stay there, Sambandar would daily go to Tiru Marugal to worship the Lord. One day a merchant had come there with his wife. When they were asleep, a poisonous snake bit the man and he died. Doctors failed to revive him. The wife prayed to the Lord for His mercy. At that time Sambandar entered the temple and heard the woman's wailing. Sambandar consoled her, and she narrated to him her story and her pitiable condition. Sambandar sang a song, and the merchant at once came back to life! All of them worshipped the saint's holy feet.

At the request of Siruthondar, Sambandar wanted to have the Darshana of the Lord at Chenkattankudi. When he was taking leave of the Lord, He gave him Darshana in the form as He is in Chenkattankudi. On the way, Sambandar stayed at Tiru Pukalur as the guest of Muruga Nayanar, and sang his glories.

At the suggestion of Appar Swamigal, Sambandar visited Tiruvarur and had Darshana of Lord Thiageesa. Then both the saints stayed with Muruga Nayanar for some time. They then went to Tiru Kadavur, met Kungiliya Kalaya Nayanar and sang his glories.

They then came to Tiruveezhimizhalai. During their stay there, the Brahmins of Sirkali met Sambandar there, and pleaded that he should go to Sirkali and have the Darshana of Lord Thoniappar. The Lord Himself, however, did not want His child to undertake this journey. The next morning, Sambandar went to the local temple for worship. There he saw Lord Thoniappar seated in front of him. He sang His glories. He informed the Bhaktas of this and sent them back to Sirkali.

Appar and Sambandar stayed at Tiruveezhimizhalai for some more time. There was a severe famine there. Appar and Sambandar were moved by the sufferings of the Bhaktas. They offered prayers to the Lord Who promised to give them some gold coins daily, with the help of which they could serve the people. Both of them found a gold coin, at different entrances to the temple. Appar was immediately able to get provisions for his gold coin, whereas Sambandar could not. He had to exchange his coins for pure gold coins, before he could obtain the provisions. Sambandar understood that it was because of Appar's sincere service to the Lord, and sang a Padigam praising the Lord. The Lord then gave him also pure gold coins and he had no difficulty in getting the foodstuffs.

The miracle that they performed at Tirumaraikadu or Vedaranyam, has already been described, while dealing with the life of Appar Swamigal.

As has already been stated, Jain influence was growing in Madura, and even the king had succumbed to it. There were only two persons who were free from the influence, and they were the queen Mangayarkarasiar and the minister Kulacchirai Nayanar. They were Shaivites by inner conviction, though they did not wear the external Saivite marks, for fear of the king's wrath. They had heard the glory of Sambandar. So, without the knowledge of the king they sent some wise men to Tirumaraikadu to persuade Sambandar to rescue Shaivism from the Jain influence. Sambandar informed Appar of his desire to leave for Madura immediately. Appar, out of sheer love for the young Boy, pleaded that he should not go, but, realising his divine nature, let him go!

The Jains living all over the Pandyan kingdom saw many evil omens. They reported to the king. At the same time, Mangayarkarasiar and Kulacchirai saw many good omens and were happy.

By this time, the news of the arrival of Sambandar reached the queen who sent the minister to welcome him. She herself went to the temple and offered special prayers to the Lord. The minister who proceeded to the border, heard the sound of trumpets and chanting of Vedas. He went towards that direction. The very sight of the Bhaktas who were coming in advance, thrilled him. He fell at their feet and did not get up at all. The devotees carried this news to Sambandar. Sambandar got down from his palanquin and went to Kulacchirai. He lifted the minister up and embraced him. Sambandar worshipped the Lord the moment he beheld the temple tower from a great distance. He sang the glories of the minister and the queen. They went to the temple. The queen, standing on one side, offered mental prostrations to Sambandar. Then, she fell at his feet. Sambandar blessed her.

The news of Sambandar's arrival had reached the Jains. The holy vibrations of the Panchakshara pierced their ears. They decided to bring the wrath of the king on those who welcomed the saint. They told him that they had all been polluted by the sight of the Shaivites who had entered the city, following the arrival of 'one young Brahmin alleged to have been blessed with Divine Knowledge by Lord Shiva directly and who wants to defeat us in a religious debate'.

The king took counsel. The Jains sought his permission to burn Sambandar's camp with the help of black magic. He gave them permission. But, it did not succeed. In the meantime, seeing the king worried, the queen ascertained the cause, and suggested that both the rival parties should be invited to argue their case and prove the superiority of their own religion. The king agreed.

The Jains failed to set fire to Sambandar's camp. So, they set fire to the camp in which the devotees were lodged. They got up, ran to Sambandar and told him what had happened. He sang a Padigam expressing the wish that (in accordance of the law of Karma) the fire for which the king was responsible should proceed towards him. Next morning, the news reached the queen and the minister. They were grieved. They wanted to put an end to their lives, but changed their mind when they heard that nothing had happened to Sambandar or the devotees.

As soon as Sambandar sang the Padigam, the fire in the camp died out and proceeded towards the king, in the form of a dreadful disease. The king experienced burning sensation all over the body. All the endeavours of the doctors and the Jain priests to alleviate the king's suffering proved futile. The queen and the ministers understood the real cause of the king's ailment and were worried. They informed the king of their feeling and requested him to call Sambandar immediately so that his grace might relieve him of the distress. The king acceded to their request and decided to embrace Sambandar's faith, if he could cure the disease.

The queen at once went out, surrounded by her maid-servants, to invite Sambandar. Kulacchiraiar also went ahead of her. They reached the Mutt in which Sambandar was staying. They fell at his feet and informed him of the king's condition: 'The atrocity of the Jains had recoiled on the king who is suffering from intense agony which the Jains have failed to relieve. With folded palms we entreat you to relieve him of the distress, and then defeat the Jains in argument and convince the king of the superiority of Shaivism.' Sambandar assured them that he would fulfil their wishes. He went to the temple to get the Lord's blessings for defeating the Jains in debate and establishing Shaivism in the land.

Followed by the queen and the minister, Sambandar went to the palace. The king had him received with all the honours.

The Jains were worried and suggested knavishly that, even if he was cured by Sambandar, he should give the credit to them only, for the preservation of Jainism! The king refused to be unjust and partial. Sambandar came into the king's apartments. The king had him seated on a nicely decorated throne, which greatly annoyed the Jains. They challenged him to a debate. The queen was afraid that they might behave in an unruly manner towards Sambandar who was but a boy in age. She suggested that the king's disease should first be cured. The king agreed to this. Sambandar also assured her that he was not afraid of anything.

The king asked the two parties to demonstrate their powers by curing his disease. The Jains volunteered to cure the disease on the left side, leaving the right to be dealt with by Sambandar. The king agreed. The Jains touched various parts of the king's body with peacock feathers, chanting their Mantras. The pain only increased! The king looked pleadingly at Sambandar. Sambandar sang a Padigam in praise of the sacred Ash (Bhasma) and with his own hand smeared the Ash on the right side of the king's body. At once the burning sensation stopped and the king experienced a cooling sensation. The king told the Jains that they had already been defeated and turned to Sambandar and entreated him to cure the disease on the left side also. Sambandar applied the holy Ash on the left side also and the disease vanished completely.

The queen and the minister fell at Sambandar's feet. The king followed suit and praised him. The Jains, however, attributed the cure to Sambandar's poetical talents, and were quite sure that he could not defeat them in philosophical arguments. They began to think of some other means of defeating Sambandar. When Sambandar invited them to open the debate, they said that they preferred practical demonstration to theoretical discussions. They wanted to challenge Sambandar to a fire test. They said that both the parties should write the essence of their respective religions on palm leaves and put them into fire: that religion should be considered as the real one whose inscriptions survived this test. Sambandar agreed to the condition.

The fire was lit. Sambandar, offering his prayers to the Lord, opened the bundle of palm leaves which contained his soul-stirring hymns on Lord Shiva and removed the Padigam which he had composed at Tiru Nallaru. To Sambandar, Lord Shiva was the Absolute Truth, and so, the song sung in praise of Him, should also be eternal. With the firm conviction that no harm would come to the palm leaf, he put it into the fire. The Jains also put their writings into the fire. The latter was at once burnt: Sambandar's leaf was quite safe. The Jains, ashamed to face the king, dropped their gaze. The king declared that the Jains had been defeated a second time.

The Jains, however, would not agree, and wanted a third test. This time both the parties should throw their palm leaves in the river Vaigai and the palm leaf which swam against the current contained the Truth. Sambandar agreed to this, too. This time Kulacchiraiyar intervened and asked: 'What should be the punishment to be meted out to the party that fails in this test?'

The Jains, in their anger, said that the party which fails in the test should be hanged. The Jains threw their palm leaf into the river: the current was swift and the leaf was washed away. Sambandar threw his leaf which swam beautifully against the current, without sinking or getting lost. In the Padigam which won this test, Sambandar invoked the Lord's grace on the king. On account of this, the king's birth-deformity, *viz.*, a hunchback was also cured.

The leaf reached the place known as Tiruvedagam. The minister wanted to take possession of the leaf and followed it. Knowing this, Sambandar sang another song, which stopped the leaf. The minister took the leaf, went to the temple and worshipped the Lord. Sambandar, accompanied by the royal couple, went to the temple and worshipped the Lord. The king was convinced of the superiority of Shaivism. The Jains, according to their own contract, were hanged. The people followed the example of the king and became Shaivites. Thus was Shaivism re-established in Madura.

In Sirkali, Sambandar's father was waiting for the illustrious son's arrival. One day, the desire was strong and Shivapada

Hridayar came to Madura and was received by Sambandar with great reverence.

After staying at Madura for some time, Sambandar proceeded on a pilgrimage, accompanied by the royal couple and the minister. From Rameswaram, they offered mental prostrations at the Feet of the Lord of Tirukonamalai and Tiruketheesvaram (in Ceylon). They also visited the birthplace of the minister. Sambandar took leave of the Pandyan king and went into the Chola kingdom.

He came to Mullivaikarai. There the river was in flood. The boatmen had abandoned their boats and had left them tied to the tree on the bank. Sambandar wanted to cross the river and worship the Lord at Tiru Kollampoothur. Sambandar asked the devotees to unfasten the boat and get into it. He sang a Padigam. This itself proved to be the oar. They reached the other side safely and worshipped the Lord.

The party then reached the place called Bodhimangai. It was a Buddhist centre. Sambandar's devotees were blowing the trumpets and singing their Guru's glories as they entered the place. This annoyed the Buddhist who asked them to stop blowing their trumpets. The devotees informed Sambandar. A disciple of Sambandar, by name Samba Saranalayar, who used to record all of Sambandar's songs, himself sang a Padigam and said that a thunder should fall on the head of Buddhanandi, the leader of the Buddhists' group. Buddhanandi was at once destroyed by a thunder.

The others fled. But, soon they reappeared under the leadership of Sari Buddhan and challenged the Shaivites to a debate. With the blessings of Sambandar, the disciple Samba Saranalayar himself defeated the Buddhists in debate. Sari Buddhan himself embraced Shaivism and his followers followed suit. Sambandar blessed them all. Sambandar then went to Tirukadavur. When he heard that Appar Swamigal was at Tirupoonthurithi, Sambandar went forward to meet him. At the same time, Appar came half-way to welcome Sambandar. Quietly, he got mixed with the crowd and joined the group of devotees who were carrying Sambandar's palanquin, without anybody's knowledge.

When Sambandar enquired about Appar, Appar responded from below: 'Here I am, carrying the palanquin, due to the virtuous deeds of many past lives.' Sambandar was surprised. He jumped down and embraced the great saint Appar.

After some more pilgrimage, Sambandar returned to Sirkali. Tiruneelakanta Yazhpanar and his wife took leave of Sambandar and returned home.

Sambandar wanted to visit Thondai Nadu. Taking leave of Lord Thoniappar, he left Sirkali, and after visiting many shrines on the way, reached Tiru Annamalai. The very sight of the hill sent him into a trance. He rolled on the ground and shed tears of God-love. Then he reached Thondai Nadu and came to Tiruvothur. During his stay there, a Bhakta came to him and said: 'I have planted many palmirah trees in my garden, but all of them are male trees and they do not yield any fruits. The Jains are mocking at me for this. Please protect me from their scorn.' Sambandar went to the temple and sang a song mentioning the devotee's plight: and the male trees were at once changed into female trees and they yielded good fruits. Due to this miracle, some more Jains embraced Shaivism. Because Sambandar had specifically mentioned the palmirah trees, they, too, were helped in their evolution.

By stages, Sambandar reached Tiru Alankadu, the holy place where Karaikkal Ammayar 'walked' on her head, not wishing to pollute the place. He, too, did not enter the place, but had the Darshana of the Lord in his dream. Sambandar then went to Kalahasthi and had the vision of Kannappa Nayanar and also of Kailash, Ketharam, Gokarnam, Tirupatham, Indraneela Parvatham, etc. Sambandar then came to Tiru-Votriyur.

In Mylapore there lived a merchant by name Shivanesar. He was a staunch Shiva Bhakta. He had all wealth but had no children. In answer to his sincere prayer, Lord Shiva blessed him with a female child. They named her Poompavai. She was very beautiful. Shivanesar heard of Sambandar's greatness and felt that he was the only suitable match for his daughter. Mentally, he had offered her to Sambandar.

One day when Poompavai was gathering flowers in the garden, she was bitten by a poisonous snake and she died. Shivanesanar even announced that he would give any amount of money to anyone who would revive her: but it was of no use. Then he recollected that he had mentally offered her to Sambandar: this put great courage into him. He at once cremated the body of the girl, collected the ashes and preserved them in a pot. Daily he would decorate the pot with flowers, etc., and sit near it meditating on Sambandar. The news that Sambandar was staying at Tiruvotriyur reached the merchant. At once he erected a big pandal from Mylapore to Tiruvotriyur and followed by Bhaktas began to proceed towards Tiruvotriyur to meet Sambandar.

The latter also was coming towards Mylapore. They met on the way. Sambandar had heard about Shivanesar and his worship of the pot which contained the ashes of his daughter. He wanted to please Shivanesar by bringing the girl back to life. They reached Mylapore, worshipped the Lord, sang hymns and, coming out of the temple, asked Shivanesanar to bring the pot of ashes. Sambandar addressed the pot: 'Oh Poompavai, the very purpose of human birth in this world is to serve the Lord and His devotees, and to feast the eyes by seeing the festivals of Lord Shiva. If this is true, arise in the presence of all. Are you going away without seeing the festival?' Then he sang a Padigam. When he finished the first stanza, Poompavai got her form. When he finished eight stanzas, she got her life and became a twelve year old girl. When he finished the tenth stanza, she came out of the pot, even as Lakshmi came out of the Lotus.

All were amazed at this miracle. Shivanesanar and Poompavai worshipped Sambandar's feet. Shivanesanar entreated Sambandar to accept the girl as his wife. Sambandar, however, explained that the original Poompavai whom Shivanesanar had mentally offered to Sambandar was dead and that the present girl had the relationship of daughter to him. Shivanesanar had to bow to the wishes of Sambandar: he built an Ashram for his daughter where she spent her days in worship of the Lord and attained Him.

After visting a number of shrines, Sambandar returned to Sirkali. He had reached his sixteenth year. His father wanted to get him married. He argued that it was necessary for him to engage himself in the performance of Vedic rites. Shivapada Hridayar selected the daughter of Nambandar Nambi of Nallur Perumanam. He, too, welcomed the alliance. The wedding was to take place at Nallur Perumanam. On the appointed day, Sambandar took leave of Thoniappar and reached Nallur Perumanam. Sambandar went to the temple, worshipped the Lord and got His blessings. Then he went to a Mutt nearby. The bride's party came there to receive him. Sambandar, in his wedding dress, took his seat in the pearl palanquin. People accompanied him, singing 'Long Live Sambandar.' Sambandar came to the place where the wedding was to take place. Sambandar holding his wife's hand, went round the fire, the manifestation of the Lord.

Accompanied by the devotees, the couple went into the temple and worshipped Him, with total self-surrender. Sambandar sang a Padigam praying for Liberation. The Lord granted his wish and said: 'Oh Sambandar, you, your wife, and all those who witnessed your marriage will merge in the Shiva Jyoti and come to Me.' At once, an effulgent Light emerged from the Lord. Before merging in that Light, Sambandar sang a Padigam known as the Panchakshara Padigam. Then all those who were there merged in the Light of Shiva. Tiruneelakanta Nayanar, Muruga Nayanar and Tiruneelakanta Yazhpanar were also there!

Eyarkon Kalikama Nayanar: In Tiruperumangalam in Ponni Nadu, there was a Vellala by name Eyarkon Kalikama Nayanar. He belonged to the family called 'Eyarkudi' which produced Commanders-in-Chief to Chola kings. He was a great devotee of Lord Shiva.

The news that Sundaramurthi Nayanar had used Lord Shiva Himself as a messenger to settle the domestic dispute between him and his wife Paravayar, so greatly annoyed devout Kalikamanar that he said to himself: 'Is this man who behaved like this a devotee? I am a great sinner, too, otherwise my life-breath would have departed on hearing the news.' Sundarar came to know of Kalikamar's attitude. He realised his own fault and entreated the Lord to appease Kalikamar's anger. The Lord caused Kalikamar to suffer from colic and told him in a dream that only Sundarar could cure it. Kalikamar preferred death to being cured by Sundarar! In the meantime, the Lord appeared in Sundarar's dream and asked him to go to Kalikamar and cure him of the colic. Sundarar at once went to meet Kalikamar. As soon as he heard that Sundarar was coming to cure him, Kalikamar gave up his life by cutting his bowel open.

Sundarar was greeted by Kalikamar's wife. When he asked for Nayanar, she made her people say that there was nothing wrong with him and that he was asleep. After much persuasion, they showed him the Nayanar's body. In desperation, Sundarar also wanted to cut his throat. By the grace of Lord Shiva, Kalikamar at once came back to life. They embraced each other. Kalikamar spent the rest of his life in the service of the Lord and His devotees and finally reached His Abode.

Tiru Mula Nayanar: Tirumula Nayanar was a Shaiva Siddha. He was one of the eight students of Tirunandi Devar Who showered His grace on them. They were all Yogis. He was called Tirumular because he entered into the mortal frame of Mulan.

Tirumular desired to see Agastya Rishi in Pothia hills. So he left Kailash and went southwards. On the way, he visited many Saivite shrines. When he came to Tiruvavaduthurai, he took bath in the river Kaveri and went to the temple. He went round the temple twice and offered prayer to the Lord. When he was walking along the bank of Kaveri, he saw a herd of cows shedding tears. He found out the cause: the cowherd lay dead. Tirumular wanted to pacify the cows. He entered the body of the cowherd after safely depositing his own body in the trunk of a tree. The cows rejoiced again. This cowherd was known as Mulan, a resident of Sattanur. In the evening, he drove the cows back into the village. Mulan's wife was eagerly expecting the return of her husband. But, when she approached him that day, he would not allow her to touch him, but said: 'Oh lady, I am not your husband. Adore Lord Shiva and attain Liberation.' He left her and went away to a near-by Math.

The lady complained to the leaders of the place, about the conduct of her husband. They examined him and came to the conclusion that he had attained great spiritual evolution. So, they asked her to leave him alone. The next day, Tirumular followed the cows, but could not find his body where he had left it. It was the Lord's Lila. Lord Shiva wanted Tirumular to write a book on Shaiva Philosophy, containing the essence of all Shiva Agamas, in Tamil. Tirumular understood His wish and returned to Tiruvavaduthurai. He worshipped the Lord and sat under the near-by peepul tree in deep meditation. He was in Samadhi for three thousand years. But, every year, he would come down from Samadhi and compose a stanza: thus, in three thousand years he wrote three thousand stanzas. This book is called Tirumandiram.

The Lord's mission had thus been fulfilled. Then, Tirumular went back to Kailash.

Dandi Adigal Nayanar: In Tiruvarur, in the Chola kingdom, there was a pious Bhakta by name Dandi Adigal. He was born blind. But, he always repeated the Panchaksharam and visualised the Lord with his inner eyes. He would daily go to the temple, do circumambulation, and worship the Lord.

There was a tank on the western side of the temple which was surrounded by Jain dwellings. Dandi Adigal wanted to extend the area of the tank. But, how could he, a man born blind, do it? But, with a determination and complete faith in the Lord, he decided to do it. He erected a post inside the tank where it was to be dug and tied a rope to it. The other end of the rope was tied to the other post which was fixed on the bank. Then guided by the rope he would go to where the digging had to be done. He would dig with a spade, collect the earth in a basket and again with the help of the rope, he would go to the bank and throw the earth away.

The Jains were watching the blind man's miracle. They were jealous of his achievement, too. So, they wanted to disturb his faith. They put forward a cunning argument. 'You are blind and you cannot see. While digging, you are killing many insects which is a great sin. So, give up this foolish act.' Nayanar explained to them the sacredness of the work, which was highly pleasing to the Lord, and also that by His grace, he was sure no insect would be injured. He went on with work. The Jains were insulting in their behaviour now. They said: 'You were born blind: but now you are deaf, too. Even though we are giving you good counsel, you are not listening.' Nayanar replied: 'Oh ignorant people, what do you know of the Lord's glories? To me He is the sole refuge. I live only to serve Him. Do not mock at His grace. By His grace, if I regained my eye-sight and you lost yours, what will you do?' This reply greatly annoyed them: they snatched the spade and the basket from his hands.

With a broken heart, Dandi Adigal went to the temple and expressed his grievances to the Lord. The Lord appeared in his dream that night and assured him of His help. He also appeared before the king in his dream and asked him to redress the grievances of Nayanar. The king summoned the Jains. Dandi Adigal was also there.

The king addressed Dandi: 'Oh devotee, you said that by God's grace you could get your eye-sight back and the Jains would lose theirs. Prove this.' Dandi Adigal said: 'Lord Shiva is the real God. He is my sole refuge and prop. I am only His slave. If this is true, let me regain my eye-sight and let the Jains lose theirs.' He then uttered the Panchaksharam and went into the tank. When he came out, his eye-sight had been restored and at the same time, the Jains lost their eye-sight. All were amazed at this. The king banished the Jains from his kingdom and restored people's faith in the Lord. Dandi Adigal attained God-realisation.

Murkha Nayanar: This saint was a Vellala by caste and he belonged to Tiruverkat in Thondai Nadu. He was a great devotee of Lord Shiva and was doing Maahesvara Puja by regularly feeding His Bhaktas at any cost. He had spent all his wealth in such feeding.

So, he resorted to a strange way. He used to gamble and use the money for feeding Shiva Bhaktas. He went away from his village in search of gamblers! He would not spare anybody: if someone refused to gamble with him, he would resort to violence! (So the name Murkha Nayanar which means wicked Nayanar!) But, he would never utilise the money for his own expenses. It was all for His Bhaktas. So, the Lord, the Indweller, showered His blessings on him.

This is an extraordinary illustration of the nature of supreme devotion or Para Bhakti. It is its own law. The devotee knows nothing but God and is actually oblivious of the world and its manners. He lives in God, for God and he is of God. At such a stage, God Himself takes charge of him! The completeness of the surrender is severely tested before this.

It is the extreme difficulty of this path that made Sage Narada exclaim that even a saint should not violate the canons of morality. Hence, so long as you are aware of your own individuality, stick to the code of right conduct: do not foolishly imitate the sages who dwell in a plane of consciousness, to which you are a complete stranger.

Somasira Nayanar: Somasira Nayanar was a Brahmin by caste. He lived in Tiruvambur. He was a great devotee of the Lord and served His Bhaktas, irrespective of their caste. He did Yagas and worshipped the Lord, without expecting any reward. He went to Tiruvarur and lived with Sundaramurthi Nayanar to whom he had totally surrendered himself. Thus he got His grace.

Here is a simple life of saintliness. On the face of it there does not seem to be anything spectacular about this Nayanar. But, we have to bear in mind the conditions that prevailed in South India in the Nayanar's days. It was almost impossible for a Brahmin in those days to mix with people of other castes, however devoted they might be to God. For a Brahmin to serve them was unthinkably difficult. It required very great will power, determination and devotion to God and His Bhaktas.

Again, in those days no one would even think of performing a Yaga without expectation of a reward. Yagas were performed only with a specific selfish desire. That Nayanar performed them selflessly and desirelessly, shows that he had already reached a high stage of Jnana or spiritual insight. He was a true Jnani and Karma Yogi.

Over and above all these, he was highly devoted to the Guru, Sundaramurthi Nayanar. What cannot Guru Bhakti achieve? And, yet, foolish and arrogant man speaks lightly of it and ridicules Guru Bhakti!

Sakkiya Nayanar: This saint was a Vellala born in Tirucchangamangai. He was totally disgusted with worldly life and wanted to attain Liberation. He sought the best way to get this. Due to false propaganda, he fell a victim to Buddhistic influence. He became a Buddhist, but it did not satisfy him for long. He was immediately attracted to Shaivism and was convinced that, whatever be the external appearance or conduct of one, if he had intense devotion to the Lord, he would attain Liberation. Though he did not give up his external appearance of a Buddhist, he adored Lord Shiva.

One day, as he was sitting in an open Shiva temple and meditating on the Lingam, completely absorbed in the divine bliss, he self-forgetfully threw a stone at the Lingam. On the next day, he went to the temple again and recollected the previous day's action. He felt that it was the Lord's Will, to reveal the profound truth that He would accept anything offered by His Bhakta in devotion. He threw a stone that day too. That was his daily worship, without which he would not take his food! One day, when he was about to take his meal, he remembered that he had not done his usual Puja: and unmindful of the hunger, went to the temple and threw a stone with great devotion. The Lord appeared before him, blessed him and took him to Kailash.

Sirappuli Nayanar: Sirappuli Nayanar was a pious Brahmin. He lived in Tiruvakur in the Chola kingdom. He was an ardent devotee of Lord Shiva and His Bhaktas. He used to worship them and serve them sincerely. He would repeat the Panchakshara Mantra, with Bhav and sincerity, throughout the day and night. He also performed the Vedic sacrifices in honour of Lord Shiva. All these earned for him the supreme grace of Shiva.

A special spiritual practice of this Nayanar seems to have been the ceaseless repetition of the Panchakshara Mantra (Om Namah Shivaya). This extremely simple practice is capable of bestowing incalculable benefit on man: and yet, ignorant man, full of delusion, refuses to resort to it. The continuous repetition of the Mantra will change the very mind-substance. It contains a divine vibration. Modern science has reached a stage when it no longer believes in mass and quantity. Even the gross and impure intellect of a scientist has come to recognise the superior power in the subtle atom or cell. Great indeed is the foolishness of man if he still refuses to believe that the sound-vibrations (even the subtler thought-vibrations) can bring about a radical change within himself and heal him physically, vitally, mentally, psychically and spiritually.

This is the highest Yoga: to remember God always and constantly to repeat His Name.

Siruthonda Nayanar: In Tiruchenkattangudi in the Chola kingdom, there lived a Shiva Bhakta by name Paranjyoti. His was a family of army commanders. He himself was the Commander-in-Chief of the Chola king. He realised that devotion to the Feet of Lord Shiva was the best means of obtaining Liberation from Samsara and so, he clung to Them.

Once, at the instance of his king, he waged war with a North Indian king, defeated him and returned with a big booty. The king was highly pleased. The minister informed the king that Paranjyotiar was able to achieve the victory because of his intense devotion to Lord Shiva. This shocked the king, who was a Shiva Bhakta himself: he regretted having compelled a Shiva Bhakta to wage a bloody war. He called Paranjyotiar, apologised for having sent him, a Shiva Bhakta to war, and, after giving him rich presents, sent him back to his village, with the request that he should henceforth engage himself in His Puja. Paranjyotiar returned to his village and from that time was engaged in the worship of the Lord and His Bhaktas. He would not eat without first feeding a Shiva Bhakta. He regarded himself as the lowly servant of the Lord and His Bhaktas: hence the name Siruthondar (small servant).

Lord Shiva wanted to bring out the glory of this noble saint. So, one day He appeared in front of Siruthondar's house, in the guise of a Vairavar (a special class of Shiva Yogis). He enquired of Siruthondar's maid-servant, Sandana Nangaiyar, whether her master was at home. She said: 'No, he has gone in search of a Shiva Bhakta, without feeding whom he would not take his food.' But, afraid lest this Shiva Yogi should go away, she entreated him to come into the house. The mendicant would not: 'I shall not enter the house in which a woman is alone.' Siruthondar's wife Tiruvengattu Nangaiyar heard these words and came out hurriedly and prayed to the Vairavar to stay in the house till the husband returned. The Vairavar repeated his objection and said: 'When he comes back tell him I am under the tree near the temple.' The Vairavar went away.

Immediately afterwards, Siruthondar returned. His wife told him all that had happened in his absence. Siruthondar was overjoyed because he was unable to find any other Bhakta that day. At once he ran to the temple and fell at the feet of the Vairavar and invited him to the house for Bhiksha. The Vairavar, however, hesitated and remarked: 'I doubt whether you will be able to fulfil the exacting conditions I shall demand for accepting your Bhiksha: so, better leave me alone.' Siruthondar was greatly grieved. He had thought that this mendicant had been specially sent by God to enable him to adhere to his vow and feed a Bhakta every day. He was prepared to meet any demand from the Bhakta, if only he agreed to take the Bhiksha. Now, the mendicant revealed his condition: 'Oh devotee, it is my habit to eat once in six months the fresh meat of a Pasu.

That time has now come. I doubt whether you will satisfy me.' This word Pasu has two meanings: an animal and a human being. Siruthondar thought that the mendicant only meant animal meat: and readily agreed! To his surprise, however, the mendicant revealed that meant human flesh! He also added: 'Oh friend, it should be the meat of a child. The child should be five years of age. He must be healthy. He should be the only son of his parents. Such a boy must be held by the mother and cut into pieces by his father. This meat must be cooked nicely and offered to me.' Without the least hesitation, Siruthondar accepted conditions and took the mendicant home.

How to find a boy of the mendicant's description? Siruthondar thought of his own son who fitted the description. The noble wife agreed, too, and asked him to get the child from school. As soon as he came the mother held him on her lap. The innocent child was laughing when Siruthondar, with one stroke cut his throat. The head is generally unfit for cooking, and is not fit for being offered to the Lord. So, they gave it away to the maid-servant and began to cook, the rest of the meat. After worshipping the mendicant, Siruthondar was preparing to offer him Bhiksha. The mendicant ascertained the method adopted by them in cooking the meat and Nayaṇar explained everything (except the fact that it was their own son that they had sacrificed). The mendicant said he would eat the head, too. The maid-servant had anticipated this and had the head cooked and ready.

Once again, Siruthondar requested the Yogi to have his meal. Now, the Yogi wanted another Shiva Bhakta to eat with him: and there was no one except the Nayanar himself. So, he sat with the Yogi, ready to eat the flesh of his son, to please the Yogi. Yet, one more condition had to be fulfilled! The Yogi said that unless the host's son ate with him, he would not eat! Nayanar tactfully explained that his son was not in the house and so could not join with them.

But, the Yogi insisted: 'Go out and call for him: he will come.' Nayanar wanted to obey the Yogi and did as the Yogi had asked to do. Wonder of wonders: the young boy came running to the father as soon as the father had cried aloud: 'Sirala, come here: the Yogi wants you to eat with him.' The parents were astonished to see their child, Siralan come back to life. They entered the kitchen, but could not find the Yogi there. The meat had also disappeared! As they were searching for the Yogi, the Lord appeared before them, blessed them and took them to His Abode.

Cheraman Perumal Nayanar: Cheraman Perumal Nayanar was born in Kodunkolur. It was the capital city of Malai Nadu or the present Kerala. He was born in the royal family of Kothayars, otherwise known as the Uthiyan family. The name Cheraman was the common name for all Cheras. Perumal was the title adopted by him after his coronation. His original name was Perum-Ma-Kothayar. He was endowed with good Samskaras. He had great devotion to the Lord even as a child. As he grew, his devotion also grew. He had a remarkable degree of dispassion and discrimination. He did not like to rule the country: and so, when he came of age, he renounced the world and went to Tiru Anchaikalam and engaged himself in the worship of the Lord there.

The country was ruled by Sengol Porayan. He, too, soon realised the evanescence of worldly life and renounced the world! He had no issues and the throne was vacant. They went to Tiru Anchaikalam and requested Perum-Ma-Kothayar to ascend the throne. Though he was reluctant, lest it should interfere with his daily worship, he bowed to the divine will. He went to the temple and offered a prayer.

The Lord permitted him to accept the rulership. By the Lord's grace he ascended the throne and ruled the country justly and wisely. He could understand all languages, even the language of the birds. The Lord had bestowed upon him all the Aiswaryas, great strength, royal vehicles, etc.

After the coronation, he went to the temple and after worshipping the Lord he was returning to the palace. On the way, he saw a washerman whose body had been smeared with white sand and mud. The very sight enraptured Cheraman who saw in him the image of Lord Shiva with the sacred ash smeared all over the body. He was raised to God-consciousness. He descended from the elephant and fell at the feet of the washerman, in spite of the latter's protest. All were wonderstruck to witness the supreme devotion of Cheraman.

By his many acts of devotion and piety, he earned the grace of Lord Shiva. The Lord sent to him a renowned musician and devotee, Banapatirar, with a palm leaf on which was the Lord's own song in praise of Cheraman! It read: 'Oh king who honours great poets with rich presents, who rules his subjects with love! Glory to you! I am very highly pleased with your devotion and charitable nature. The bearer of this message is Banapatirar who is a great devotee like you. He is a great musician and always sings My glories on his favourite instrument, Yazh. He has come to see you.

Gananatha Nayanar: Gananathar was a pious Brahmin of Sirkali. He was a great Bhakta of Lord Shiva. People admired his virtue and devotion and came to him for advice. He invariably gave them some work connected with the temple, according to their ability. They would clean the temple, make garlands, work in the gardens, burn lamps in the temples, etc. Thus he infused devotion in them for the Lord and transformed them into Shiva Bhaktas. He was greatly devoted to Jnana Sambandar. All these earned Shiva's grace for him.

Here is another great but simple spiritual practice. Talk to other people about God and the glory of devotion, etc. You will be building up a powerful spiritual fortress around you. You will be able to avoid people wasting your time: you will not indulge in nor allow others to lead you into, gossip which is the spiritual aspirant's arch-enemy.

People who may in the beginning, think that you are strange in your behaviour, will soon understand you and they will, of their own accord, avoid useless talk in your very presence. At the same time, you will be rendering a very great service to humanity, by directing everyone's mind towards God and Dharma. Here is a wonderful Yoga which helps you and others, too, at the same time. Put it into practice and realise its miraculous effect.

Kootruva Nayanar: This saint was a chieftain of Kalandai. He was extremely devoted to Lord Shiva. Daily he would repeat the Panchakshara Mantra, and serve the Bhaktas. The Lord made him more powerful, and bestowed on him all wealth and strength. Nayanar captured many places in the Chola and Pandyan kingdoms. He wanted that the three thousand Brahmins of Tillai should crown him king: they refused, because they owed allegiance to the Chola king. All of them even left Tillai fearing his wrath. Only one was left to do the Puja. Even he refused to accede to Nayanar's request. One day Nayanar prayed to the Lord: and, in answer, the Lord Himself appeared before him and crowned him by placing His Feet on his head. Nayanar continued to worship the Lord and finally attained Him.

The Lord, the Indweller of our hearts knew that, when the Nayanar asked the three thousand Brahmins to crown him, it was only to spiritualise the coronation and to enable him to feel that the crown was but a symbol of the Feet of the Lord. When the Brahmins feared political repercussions, the Lord Himself fulfilled the devotee's wish.

Pugal Chola Nayanar: Pugal Chola Nayanar was a king. He was living in Uraiyur in the Chola kingdom. He was greatly devoted to Lord Shiva and His Bhaktas. He was an ideal king and people loved him and followed in his footsteps.

Once he went to Karur to collect tributes due to him from the kings of Kuda Nadu. All of them paid at once: but the ministers reported that a petty king named Adigan had not. He ordered his troops to invade Adigan's fort. In the meantime, the king's elephant was killed by Eripatha Nayanar for a Shiva Aparadham (as we have already seen in Eripatha Nayanar's life). Ultimately, both Eripathar and the king had the Lord's Darshana. As this drama was being enacted, elsewhere, the king's troops had demolished Adigan's fort, killing many of his men, and Adigan himself had run away. Pugal Cholar's troops returned with a lot of wealth and the heads of men killed. They placed all these at the king's feet. Among the heads, the king noticed a head with the braid of hair on top—it belonged to a Shiva Bhakta. Stricken with terrible remorse the king had a big fire made, went round it having the head on a golden plate in his hand and entered the fire chanting the Panchakshara Mantra. Thus he entered the Lord's Abode.

Narasinga Muniyaraiyar: This saint was a petty chieftain. He lived in Tiru Munaipadi. He was highly devoted to Lord Shiva. He was a champion of Shaivism. On every Tiruvathirai day he would conduct special Puja, feed Shiva Bhaktas in whatever form they appeared and offer the gift of a hundred gold coins to each. On one such occasion, one Bhakta came with sacred ashes on his stark naked body: this evoked disgust in the hearts of the other Bhaktas. Nayanar understood this and fell at the naked devotee's feet and welcomed him with more respect. He fed him nicely and gave him 200 gold coins. This earned the Lord's supreme grace for the Nayanar.

Here is an illustration of the subtle way in which saints manifest their cosmic vision, and also the subtle way in which they bring about the necessary change in the outlook of others. To the Nayanar, all the devotees are the manifestations of Lord Shiva. The naked man does not evoke the least trace of disgust or contempt. When he finds this unhealthy attitude in others, he does not violently correct them. In his own subtle, mysterious but very effective way, he demonstrates the truth: and brings about a change in the attitude of the ignorant. Both these lessons are important.

Adipattha Nayanar: This saint was a fisherman born in Nulaipadi near Nagapattinam. It was his practice to let go one fish from his catch daily, as an offering unto the Lord. The Lord wanted to reveal his greatness to the world. Once it so happened that for many consecutive days he could catch only one fish. He let it go, in the name of Lord Shiva, and went without food. One day he caught a golden fish, again only one for the day. And, he stuck to his vow and let it go, in the name of Lord Shiva. The Lord appeared before him and blessed this illiterate, fisherman saint!

Not indeed by vast erudition, nor by breath-taking austerities, nor by hearing and talking a lot, but by unflinching devotion alone can God be realised. This humble, simple, fisherman saint has proved that beyond the least trace of doubt. But, look at his steadfastness, Nishta! It is not easy to acquire, unless you have living faith in God. Otherwise, the mind will bring up all sorts of reasons (lame excuses!) for breaking the vow.

This supreme faith and devotion is itself the highest Jnana. Only an ignorant man studies books: what need is there for a great scholar to study an elementary book on grammar? What need is there for one to whom God is a living presence, to stuff himself with words? Intellect is a help, if it serves faith: it is a hindrance if it shakes it. Devotion is indispensable for attaining Him.

Kalikamba Nayanar: This saint was a Vaisya by caste and he lived in Pennagadam in the Chola kingdom. He was doing some service in the temple. He was very devoted to Shiva Bhaktas in whom he saw the Lord Himself. His wife, too, helped him in all this.

One day, his former servant came to his house in the guise of a Bhakta. Nayanar, as usual, welcomed him, washed his feet and worshipped him. But, his wife who recognised the former servant did not join. Nayanar understood her lack of devotion, cut off her hand and continued his worship of the Bhakta. Practising this form of Yoga Sadhana, he attained Him.

This saint has dramatically proved the great truth which Sage Narada has emphasised in his immortal work 'Bhakti Sutras': there is no distinction of caste, creed or status, among the devotees of God. They do not recognise such distinctions among themselves: and others, too, shall not entertain such distinctions. It is blasphemy against God. The great devotees have become one with Him. They are all manifest divinity. Whatever might have been their past life, caste or faith, they have now become divine and hence such distinctions have no meaning.

Kalia Nayanar: Kalia Nayanar was an oil monger of Tiruvotriyur. His adoration of the Lord, to Whom he was highly devoted, took the form of lighting the temple lamps daily. He was rich. But, in order to reveal his greatness the Lord made him poor. His family people also refused help. He began to work as a labourer to earn the oil. Even this became impossible. He wanted to sell his wife: but no one would buy. At last, in despair, he wanted to cut his own throat and use the blood instead of oil, to burn the lamps. In that attempt, Lord Shiva caught hold of his hand and blessed him.

What greatness, and what intensity of devotion is portrayed in this simple life! Self-forgetfulness is the key-note in devotion. Remembering God always, the devotee is so thoroughly absorbed in Him, that nothing but God and His worship matters to him. By all means His worship must go on: no obstacle shall stand in the way. The devotee's heart and mind are always positive, never letting a negative thought enter them. He sees opportunities in difficulties and is never beaten by any obstacles which serve him as steps to God!

Satti Nayanar: Satti Nayanar was a Vellala by caste. He was born in Varinjiyur in the Chola kingdom. He was a sincere devotee of Lord Shiva and honoured His devotees. He could not tolerate anyone speaking ill of them. If anyone did so, he would cut off the slanderer's tongue. Lord Shiva understood his pure inner Bhav and showered His grace on him.

Besides revealing the glory of the Nayanar's devotion, this simple life also holds for us a great object lesson—never speak ill of the saints or devotees of God. They have attained union with God: and, so, if you vilify them, you are vilifying God Himself. It is the greatest sin, the greatest, Himalayan blunder. You cannot judge them: they live on a different plane of consciousness from yours.

Our scriptures contain numerous illustrations of the strange behaviour of saints, sages and Yogis. Sometimes they behave as little children: sometimes as mad-men; sometimes as fools. Mysterious is the nature of saints. Always worship and adore them: you will be benefited. Do not criticise them or speak ill of them or find fault with their conduct. Our scriptures say that he who blames the conduct of the sages, gets their bad Karma, and suffers doubly in consequence. Beware!

If anyone speaks ill of a saint or devotee of God, in your presence, leave that place at once. Otherwise, your own moral and spiritual structure will be dangerously undermined. Beware!

Aiyadigal Kadavarkon Nayanar: Aiyadigal Kadavarkon Nayanar was a Pallava king who ruled over Kanchi. He infused the spirit of Shaivism into his people, too.

Soon he got disgusted with worldly life, renounced the world, after installing his son in his place, and undertook a continuous pilgrimage of the various shrines singing hymns in His praise, wherever he went. Lord Shiva was highly pleased with his devotion and blessed him with His Darshana.

This king has set an example for all kings and social leaders to follow. Leaders should be the best example for their followers. They should encourage the masses to walk the path of virtue and Godlines. Otherwise, they live in vain: and, what is more, they take upon their shoulders a good part of the sins of their followers, for they are responsible for those sins.

This Nayanar lives in our hearts even today because he served his subjects both by precept and by his own personal example. He taught them; he encouraged them; and, finally, by his own life of renunciation and wholehearted devotion of the Lord, set a glorious example for them to emulate. Such is the life of an ideal leader.

Kanampulla Nayanar: Kanampulla Nayanar was a wealthy man in Irukkuvelur. He was a great Shiva Bhakta. He wanted to utilise all his wealth in His service only. So, with unswerving devotion he would light the lamps in Shiva shrines and sing His praise. Lord Shiva wanted to reveal his devotion. He withdrew his wealth. He went over to Chidambaran. There also he continued his service, with the money got by selling his possessions. There was nothing left in the house. He had to cut grass, sell it and purchase ghee with the money and burn the lamps. Because he cut grass known as Kanampul, he was known as Kanampulla Nayanar. One day he could not sell the grass. He did not want to swerve from his duty, however. He went to the temple and made a wick out of the grass and burnt it. The quantity of grass was not enough. So, he brought his own head near the lamp, spread his hair on the lamp, and began burning it. At once Lord Shiva appeared before him, and blessed him.

Kari Nayanar: This ardent devotee of Lord Shiva was a native of Tirukadavur. He was a scholar in Tamil. He used to go to the three Tamil kings, and get money by singing Tamil Kovais (anthology). He earned a lot and built temples. Thus he spread the cause of Shaivism. He also served Shiva Bhaktas and earned His grace.

A true devotee of the Lord lives for His sake only. All that he has is offered to the Lord as the devotee's worship. God is your creator, your father and mother, friend and Guru. He has given you the various talents and skill. They belong to Him. He dwells in all, and expects you to utilise them in the service of all. This is the simple logic by which the saints arrive at the conclusion that they should see God in all, serve the Lord in all, and love Him in and through cosmic love, expressed as selfless service and charity.

By leading such a selfless and divine life, you will conquer your worst spiritual enemy, *viz.*, egoism, and realise God in this very birth, nay, this very second.

Mangayarkarasiyar: Koon Pandyan, the Pandyan king, was ruling in Madura. He was called Koon Pandyan because of his hunchback. He was himself a poet and he patronised the Tamil poets and established a Tamil Sangam. His wife was Mangayarkarasiyar. She was the daughter of a Chola king. She was an ardent devotee of Lord Shiva. Kulacchirai Nayanar was his minister: and he was also a staunch devotee of Lord Shiva. Tiru Jnana Sambandar has sung Padigams in praise of both.

Koon Pandyan had fallen a victim to the influence of Jainism. The queen and the minister feared that unless something was done, Shaivism would be wiped out. When Sambandar came to Madura and was staying outside the city, Kulacchirai Nayanar invited him into the city. The Jains tried in vain to destroy Sambandar. When Sambandar sang a song, the king's hunchback was cured, as also his burning pain. He came back to Shaivism. Since then he was known as Ninra Seer Nadumara Nayanar, as his hunchback had disappeared and he stood erect and tall.

The Pandyan king then defeated the northern kings at Tirunelvely and spread Shaivism there. Mangayarkarasiyar helped her husband a lot in this. Both the husband and the wife worshipped Sambandar with great faith and devotion. Their devotion to the Guru and love of Shaivism earned His grace for them.

Vayilar Nayanar: He was a Vellala by caste. He belonged to Mylapore. He was a Shiva Bhakta. He constructed temples mentally and did Manasic (mental) worship. He built the temple of non-forgetfulness, lit the shining lamp of Self-illumination, bathed the Lord in the waters of immortal Ananda (bliss) and worshipped Him with the elixir of supreme love. Thus he obtained salvation.

Here is the life of a Para Bhakta, a supreme devotee. He had transcended the stage of idol worship. He had attained great purity of heart and clarity of inner psychic vision so that, without the aid of a symbol and without the help of rituals, he could raise his mind to the sublime heights of the Abstract.

The inclusion of this wonderfully simple life of Vayilar Nayanar is to point out that devotion is of many types, to suit the taste and temperament of different individuals. Whatever be the path the choose, ultimately they reach the same goal, union with the Lord, Shiva. The Hindu sages have always declared that the spiritual path is not a stereotyped one, the same drug for all diseases, the same food for all people at all ages (from infancy to old age!), but that the spiritual life is adapted (within broad limits) to the needs of each individual. Everyone pursues the path or the combination of paths suited to him, and ultimately reaches the same goal.

Munaiyaduvar Nayanar: This saint was a Vellala by caste. He belonged to Tiru Nidur in the Chola kingdom. He was a great Bhakta of Lord Shiva and His devotees. He was always the hope of the desperate, the weak and the vanquished. They would call upon him to turn their defeat into a victory. He would hire himself out as a professional fighter. He fixed a wage for this service and with that money he would feed the Shiva Bhaktas and look after them. He earned money in this way and hence he was called 'Munaiyaduvar'. Lord Shiva was highly pleased with him and blessed him.

Two vital lessons that this Nayanar's life hold should not be ignored. The first and foremost, even in the exercise of the God-given talent of fencing, the Nayanar took care to see, that it was used to defend the weak, the oppressed and the downtrodden. Strength, too, is a manifestation of the Lord, according to Him: but it should be used in His service in a righteous way. The second one is that the fruits of such service were always dedicated to the Lord. This is the very core of the teaching of the Bhagwad Gita, and the teachings of all saints and sages. Righteousness rests on this pedestal of dedication to God and unselfishness. Selfishness is the root cause of all sins and consequent miseries.

Seruthunai Nayanar: Kazharsinga Nayanar was an ardent devotee of the Lord. He was a Pallava monarch, belonging to the family of Kadavar. Due to God's grace he defeated the kings of the northern country and established Shaivism there. He went on many pilgrimages. Once he came to Tiruvarur with his queen and visited the temple. The queen, coming round the temple, came to the place where flowers had been kept for Shiva's worship, and she smelt a flower which had accidentally fallen on the floor. Seruthunai Nayanar, a pious Vellala of Tanjore, who was doing the service in the temple, was annoyed by her action. He at once cut off the nose of the queen that smelt the flower.

The king, hearing the pitiable cry of the queen, rushed to the spot. He was terribly angry with the man who was responsible for the brutal act. Seruthunai Nayanar explained to him the queen's action which was an insult to Lord Shiva (Shiva Aparạdham). The king at once gave an additional punishment to her, by cutting off her hand which picked up the flower! Both the king and Seruthunai Nayanar were glorified by the people and the celestials rained flowers on them. Both of them attained the grace of Lord Shiva.

Idangazhi Nayanar: This saint was the king of Velas in Kodumbalur. He was a staunch devotee of Lord Shiva. He had made arrangements with all the Shiva temples to perform worship according to the Shiva Agamas. There was another Shiva Bhakta in the same locality doing Maaheshwara Puja. He became very poor and so he could not continue his Puja and feeding of Bhaktas. So, one day he entered Idangazhi Nayanar's granary at night and began to steal paddy. The watchman caught him red-handed and took him to the king. The king learnt on enquiry that the Shiva Bhakta's motive for stealing was to feed the devotees of the Lord. The king let him go.

This incident opened the eyes of the king. He realised that nothing belonged to him and that the real owners of his property were Lord Shiva and His Bhaktas. So, he gave permission to all Shiva Bhaktas to enter his palace and granary and take whatever they wanted. Thus he displayed his zeal for the spread of Shaivism. Thus he earned Lord Shiva's grace, too.

Pugazh Tunai Nayanar: This saint was a pious Adi Shaiva of Seruviliputhur. He was an ardent Shiva Bhakta. He was a Pujari (priest) in the temple. His daily duty was to bathe the image, uttering the Mantras and do the Puja, according to the Shiva Agamas. Once a famine swept over the land and he had no money to buy food. He loved the deity and his daily duty so much that he did not like to leave the place in spite of the starvation. He stuck to that place and continued the Puja. His body was emaciated.

One day, in spite of his weakness, he fetched water for the Lord's Bath (Abhishekam) and, when he was pouring the water on the Lingam, the water-pot slipped from his hand and fell on Him. Nayanar forgot himself in sheer exhaustion and fainted away. The Lord appeared in his dream and said that He would leave one coin in the temple every day till the famine was over so that he could procure the necessary food with that money and appease his hunger. Nayanar woke up and found that the dream was true! The Lord thus enabled His Bhakta to get over the famine. He continued his daily Puja in the temple and finally reached the Lord's Abode.

Kotpuli Nayanar: This saint was a Vellala by caste. He was the Commander-in-Chief of a Chola king. He was highly devoted to Lord Shiva. He was very pious and virtuous. It was his practice to purchase paddy out of his income and give it to Shiva temples for the Lord's food. He was doing this for a long time.

Once he had to go out on military duty. So, he stocked a sufficient quantity of paddy for the temple use, handed it over to his relatives, with clear instructions that it was meant only for the Lord and that they should not touch it for their own use. During his absence, there was a famine and his relatives had to suffer for want of food. So, they laid their hands on the paddy meant for the Lord and appeased their hunger. The Nayanar returned from his duty and heard of his relatives' action. He was annoyed with them. He called them to his house and killed them, including his parents, for this crime. His supreme love for the Lord had so completely overshadowed his love for his own near and dear ones! The Lord appeared at once before him and blessed him, and also all the relatives who had died at his hand, and took them all to His Abode.

Pusalar Nayanar: Pusalar was a Brahmin of Tiru Ninravur in Thondai Mandalam. He excelled in the mental worship of the Lord. Mental worship is thousands of times better than external ritualistic worship: mental worship soon leads to Samadhi (superconscious state) and Self-realisation. He strongly desired to build a temple for Lord Shiva, but he did not have the money for it. So, mentally he gathered the necessary materials for the purpose. He laid the foundation stone on an auspicious day. He raised the temple and had even fixed an auspicious day for the installation of the deity in it.

The Kadava king who was also a great devotee of Lord Shiva had built a magnificent temple in Conjeevaram. By chance he had also fixed the date which Pusalar had mentally chosen, for the installation of the Lord in his temple. The Lord wanted to show the king the superiority of Pusalar's great devotion. So, the Lord appeared in the king's dream and asked him to postpone the installation ceremony in his temple, as He would be going to the temple constructed by His devotee at Tiru Ninravur. The king woke up from sleep and was intensely eager to have the Darshana of the devotee mentioned by the Lord and also have a look at the great temple he had built, which he thought would be far superior to his temple.

The king came to Tiruninravur and searched all over the place for the temple: he could not find any. Then the king enquired about Pusalar. He found out Pusalar's house and approached him. Pusalar was stunned when he heard of the king's dream. Soon, he recovered and was filled with joy. He thought: 'How kind and merciful is the Lord. I am only a wretched creature and He has accepted my mental shrine as His Abode. I am really blessed.' He told the king that that temple was only in his mind. The king was greatly surprised to hear this. Admiring Pulasar's devotion, the king fell at his feet and worshipped him. Pulasar installed the Lord in his mental temple and continued to worship Him till he attained His Abode.

Nesa Nayanar: This saint was the native of Kampili. He was a weaver. He was highly devoted to the Lord and His Bhaktas. His mind was well fixed on the lotus feet of the Lord.

His lips always uttered the Panchakshara Mantra. His hands were ever busy in the service of His Bhaktas. These three virtues gained the Lord's grace for him.

Here is another instance of the glory of the Name of God. We have already seen the glory of the Lord's Name (Vide page 51) while studying the life of Sirapulli Nayanar. Constant repetition of the Mantra enables you to remember Him always, throughout the day and even during sleep! The technique is this: as soon as you wake up in the morning, sit down for half an hour and mentally repeat the Mantra. And keep up the current during your work, too, by withdrawing yourself for a few moments every hour and mentally visualising the presence of the Lord in you and mentally repeating the Mantra. If you are established in this practice, very soon you will find that even when you are talking or are engaged in other activities—nay, even during sleep, the mind goes on repeating the Mantra. You will get God-realisation. In addition to this glorious Japa Yoga, Nesa Nayanar also practised the Yoga of Synthesis. He thought of God, he lived for God, he worked for God, he was highly devoted to God and loved Him.

Kochengat Chola Nayanar: In Chandra Tirtha in the Chola kingdom there was a thick grove. In that grove under a Jambul tree there was a Shiva Lingam. A white elephant used to come there daily and prostrate before the Lingam. A spider which was also devoted to Him, noticed that dry leaves were falling on Him and to prevent this wove a web above the Lingam.

The next day when the elephant came to worship, he found the web, and, thinking that someone had polluted the place, tore the web, offered his worship and went away. The spider came upon the scene, felt sorry that his web had been destroyed, wove another web and went away. The next day, as the elephant was pulling the web away, the spider which was present there, gave him a sting: the elephant died of the poison on the spot. The spider, too, was caught in the elephant's trunk, and perished.

Due to His grace, this spider was born as the son of Suba Devan, the Chola king. He and his dutiful wife went to Chidambaram and eagerly prayed to the Lord Nataraja for a son. The Lord granted their wish. Soon Kamalavati conceived the child. The day of delivery arrived. Astrologers foretold that if the child could be delivered a few minutes later, it would rule the three worlds!

The queen asked that she should be tied to the roof of the room upside down, with a tight bandage around her waist. When the auspicious time came, she was released and the child was born. This was the spider reborn! The child had red eyes as he had remained in his mother's womb a little longer. The mother, looking into his eyes, said: 'Kochekannano' (the child with red eyes), and expired. Hence, he was named Kochengat Cholan. When he reached the proper age, his father enthroned him king, retired from the world and, after severe penance, reached the Lord's Abode.

Kochengat Cholan promoted Shaivism. In Tiru Anai Ka he built a beautiful temple and installed the Shiva Lingam under the same Jambul tree! In Chola Nadu he built many shrines and mansions for the use of the three thousand Brahmins of Tillai. He provided for regular worship at Chidambaran. Finally he reached the Lord's Abode. His glories were sung by the poet Poygayar in his 'Kalavazhi Narpathu'.

Tiru Neelakantha Yazhpanar: In Tiru Erukattanpuliyur, in the Chola kingdom, there lived an ardent devotee of Lord Shiva by name Tiru Neelakantha Yazhpanar. He was an expert in playing the Yazh (Veena, a musical instrument). It was his habit to visit many sacred shrines and sing His glories on the Yazh. He once went to Madura. He was standing at the entrance and singing.

The Lord wanted to hear him at close quarters and so asked the devotees in their dream, to bring Yazhpanar into the inner shrine the next day. When the Brahmins took him inside the shrine, Yazhpanar was surprised, but understood it was His Lila and that He wanted to hear him play on the Yazh. As he was singing, a voice was heard in the heaven: 'If the instrument rests on the wet floor, it will be spoilt: give him a golden seat to occupy.' At once a golden seat was offered to him. Yazhpanar prostrated to the Lord and sang of His supreme compassion, standing on the golden seat.

Yazhpanar then went to Tiruvarur and, here, too, he remained outside the shrine and sang. And here, too, the Lord wanted him to sing in His immediate presence. So, He created another opening on the northern side of the temple. Yazhpanar understood the Lord's will and entered through the gate and sang in His Presence. How he joined Sambandar and got Liberation, has been told in Sambandar's life.

Isaijnaniyar: In Tirunavalur in Tirumuraipadi there lived an Adi Shaivite by name Sadayanar. All his ancestors were ardent devotees of Lord Shiva. He was also pious and devoted. Isaijnaniyar was his dutiful wife. She was also devoted to the Lord. Due to their virtuous deeds in their past life, a divine child was born to them. He was no other than Sundaramurthi Nayanar. Narasinga Munaiyar, the king, was attracted by the child's beauty and wanted to bring it up himself. The king approached the parents and they, without a moment's hesitation, handed the child over to him. By this action, they showed that they had no attachment at all to anything in this world.

They led the ideal Grihastha (household) life and finally attained His grace. Worldly attachment is the only chain with which man binds himself to this Samsara. When there is attachment, there is Samsara or bondage: if you are completely detached, you are at once freed, you become a Jivanmukta. You enjoy the Bliss of Brahman here and now, this very moment. This is the unique glory of Hinduism: it promises immediate Liberation here in this world, while yet embodied! Immediately the entire world is transformed into a manifestation of Divine Light. All the paradoxes and mysteries of Creation are understood.

Theyyam

The Theyyam or Theyyattam is a popular ritual dance of north Kerala, south India, particularly presented in the Kolathunadu (of the present Kannur and Kasargode districts). As a living cult with centuries old traditions, ritual and custom, it embraces almost all castes and classes of Hindu religion in this region. The term Theyyam is a corrupt form of daivam or

God. It is a rare combination of dance and music and reflects important features of a tribal culture.

History

The earliest Brahminic settlements like Payyanur and Perimchellur (Thaliparamba) in Kolathunadu where the Brahminic religion was propagated through the institutions of temples largely influenced the popular folk religion based on Theyyam and other tribal cults. According to the legendary Keralolpathi, Parasurama sanctioned the festivals like Kaliyattam, Puravela and Deivattam or Theyyattam to the people of Kerala. He assigned the responsibility of Theyyam dance to the indigenous communities like Panan, Velan and Vannan. These traditions explain how the indigenous cults like Theyyam were incorporated and metamorphosed under the religious supremacy of the Brahmanism.

In the long historical process a social system evolved in Kerala in which the little culture like Theyyam belonged to the depressed castes and classes where as the temple oriented culture belonged to the dominant castes and classes. There were no violent confrontations between these two cultures as there was no total destruction of the indigenous culture. *"There can be no doubt", say Bridget and Raymond Alchin, 'that a very large part of this modern folk religion is extremely ancient and contains traits which originated ruing the earliest periods of Neolithic, Chalcolithic settlement and expression* (The Birth of Indian Civilization 1968 p.3039).

Classification of Sub Cults

Under the impact of Aryan religions, the cult of Theyyam had changed substantially incorporating new trends and sub cults along with its tribal character.

In a different way, it can be stated that all prominent characteristics of a primitive tribal religious worship had widened the stream of Theyyam cult and made it a deep rooted folk religion of the millions. For instance, the cult of Mother Goddesses had an important place in Theyyam. Besides this, the practice like spirit-worship, hero-worship, masathi-worship, tree-worship, ancestor-worship, animal worship, serpent-worship, worship of the goddesses of disease and gramaadevata-worship are included in the main stream of Theyyam cult. Under the influence of Aryan myths and legends, a large number of Brahminical gods and goddesses had infiltrated as separate cults into Theyyam. Along with these gods and goddesses there exist innumerable folk gods and goddesses. Most of these goddesses are known as Bhagavathis as a matter of sanskritisation.

Different branches of Brahminical religion such as Shaktism, Vaishnavism and Shaivism now dominate the cult of Theyyam. However the forms of propitiation and other rituals are continuation of a historical past. In several cult-centers, blood offering is forbidden under the influence of Buddhism, Jainism and Brahmanism. In such centres separate places outside the outer wall of the shrine are selected for blood offering and preparation of the traditional kalam known as vatakkanvathil. The Theyyam deities propitiated through cock-sacrifice will not enter inside such shrines walls. This is a good example of a long-standing cultural synthesis of 'little' and 'great' cultures.

On account of the later origin of Vaishnavism in Kerala, it has no widespread influence on Theyyam cult. Only a few deities are available under this category. This may probably be due to the lesser influence of Vishnu on the village folk who had an uninterrupted tradition of the worship of Mother goddess for fertility and the god Shiva and his son Murukan for protection and security even during the Sangam age. Two major Theyyam deities of Vaishnavism are Vishnumoorthi and Daivathar. Vaishnavism was very popular in Tuluva country during the 13th century under Vishnuvardhana of Hoysalas. He was a great champion of Vaishnavism. Most probably he was deified as Vishnumoorthi and propitiated in the Bhoota cult of Tuluva and then propitiated as a prominent folk deity in the Theyyam also. The legend of Vishnumoorthi is identified the God's migration from Mangalore to Kolathunadu.

All other categories of Theyyam deities can be incorporated in Shaivism and Shaktism. Even ancestors, heroes, animals etc are deified and included in those categories. In brief Theyyam provides a good example for the religious evolution and its different stages in Hinduism.

Patronage by Brahmins

When the cult of Theyyam borrowed liberally from Brahmanism or it was super imposed by the Brahmanical legends and myths, the Brahmins with their social and caste superiority, also patronized the Theyyam gods and goddesses.

They even established their own shrines and kavus (groves) for Theyyam deities where non-brahmanical rituals and customs are observed. The goddesses like Rakteshwari, Chamundi, Someshwari, Kurathi, and the gods like Vishnumoorthi are propitiated in these house-hold shrines. There, the Theyyam dancers appear during the annual festivals of gods and goddesses. The rituals in such shrines are different from those of the Brahmanical temples. Such a cultural fusion or interaction between the 'little' and 'great' cultures makes Theyyam an interesting field of research for social scientists. The impact of this cultural fusion could be traced on social organization based on caste system and in the agrarian relations.

Once the cult is patronized by the Brahmins, the intermediary and lower castes also took it as a major religious practice. In fact the cult has become the religion of the masses. As a matter of fact there is anther school of thought which believes that theyyam dances had their origin as a result of brahmins not permitting the other castes to enter temples which generally were managed by the brahmins. Theyyam emerged as a form of worshipping the same gods *i.e.* is Shiva, Vishnu and the Mother Goddesss but under different names.

A Paragraph on Performance

The dance or invocation is generally performed in front of the village shrines. It is also performed in the houses as ancestor worship with elaborate rite and rituals.

There is no stage or curtain and other arrangements for the performance. The devotees would be standing or some of them would be sitting on a sacred tree in front of the shrine. In brief it is an open theatre. A performance of a particular deity according to its significance and hierarchy in the shrine continues for 12 to 24 hours with intervals. The chief dancer who propitiates the central deity of the shrine has to reside in the rituals. This may be an impact of Jainism and Buddhism. Further after sun set this particular dancer would not eat anything as legacy of Jainism. His make-up is done by specialists and other dancers. First part of the performance is usually known as vellattam or thottam. It is performed without proper make-up or decorative costume. Only a small red headdress is worn on this occasion.

The dancer along with drummers recites the particular ritual song, which describes the myths and legends of the particular ritual song, which describes the myths, and legends of the particular deity of the shrine or the folk deity to be propitiated. This is accompanied by the playing of folk musical instruments. After finishing this primary ritualistic part of the invocation the dancer returns to the green room. Again after a short interval he appears with proper make-up and costumes. There are different patterns of face-painting. Some of these patterns are called vairadelam, kattaram, kozhipuspam, kotumpurikam, and prakkezhuthu. Mostly primary and secondary colours are applied with contrast for face painting. It had effected certain stylization also.

Then the dancer comes in front of the shrine and gradually "metamorphosises" as the particular deity of the shrine. He, after observation of certain rituals places the head-dress on his head and dances. In the background folk musical instruments like chenda, tuti, kuzhal and veekni are played with rhythm. All dancers take a shield and kadthala (sword) in their hands as continuation of the cult of weapon. Then the dancer circumambulates the shrine, runs in the courtyard dances.

The Theyyam dance has different steps known as kalaasams. Each kalaasam is repeated systematically from first to eight step of footwork. A performance is a combination of playing of musical instruments, vocal recitation, dance and strange makeup and costumes. The stage-practices of Theyyam and its ritualistic observations make it one of the fascinating theatrical arts of India.

3

Shiva-nataraja : The Lord of the Dance and Drama

Shiva as the cosmic dancer, as transformer, the liberator of energy from the spiralling and coiling nature, primordial, mysterious, symbol of spiritual ardour and passion for accomplishment and continuity of life, as the regenerator of the universe is the most classic and revered, well-known icon of Hinduism. The movements of Shiva's dance orders the rhythms and melodies of the universe, at once a dance into meditative states of transcendence. His dance is the dance of cosmic evolution of matter which knows no beginning, no middle, and no end, and is entwined and engulfed in spirit–from cell to spirituality.

In consonance with the new age science and theoretical energy-physics, matter has transcended the process of showing connectedness with the whole and its interconnectedness within and without, with spirit, consciousness and energy moving in continual dance of transformation. Many parallels are found in Shiva's dance representing the endless cycles of life, as they move from state to state, from morphose to metamorphoses. A new world view is emergent–one that is related closely to the views of mystics of all ages and traditions, and subtle energy medicine and science.

Shiva is known as the destroyer of evil and ignorance, not destroyer for the sake of destruction, but for the sake of regeneration; transformation, transmutation, for recreating, cleansing out and eliminating the debris, the dross and the gross, transforming into strength, power, force (Shiva tramples the dwarf of ignorance and destroys Muyalakan, demons).

Om Namo Shivayah as mantra represents fearlessness and protection against evil, accident, and death. Its philosophy of involution and evolution of energy and spirit leaves its imprint in every material, physical, fiber, nerve, cell, atom, molecule, proton, photon – the power to transcend – from the individual to the collectivity. In the words of Lama Angarika Govinda "to the enlightened man (human) whose consciousness embraces the universe, to him (her) the universe becomes his/her body".

In the yoga of the ida and pingala nadi blending into Sushumna from the physical, vital, mental and supramental, to infusing and permeating power of the spirit into matter, in the dynamic Tandava Nritya – dance of bliss – the body becomes the spirit, and explicates the ultimate principle of Beauty–Sundaram–in bhakti yoga, in devotion, purity, piety (in puja ritual) in integrating it with the Eternal Moral Order *i.e.* applying the principle of Ultimate Goodness—in the movement of the wheel of Karma and Dharma–Truth and Righteousness as applied to individual and collective life and order of humanity.

Shiva depicted as Ardhanarishvara–half man and half woman represents the union and differentiation of the Yin and Yang principle, the eternal feminine and masculine principle, complementing, supplementing, initiating and completing and fulfilling the divine play of the Transcendent Reality with the manifest reality in the phenomenal world of myriad changes of names and forms in lila (diverse, divine play) and unique rhythms and melodies of Shiva's energies in the cosmos. Thus manifesting Satyam–Truth, Shivam–Goodness and Righteousness, and Sundaram–Beauty as the supportive and connecting principle in the universe, spelling out the principle of Love that integrates, unites and yet transcends all threads, beads and deeds into dynamic action for realization of the Supreme.

Shiva's counterparts or divine consorts are known as Shakti, Parvati, Bhavani, Uma, Maheshevari, et al. represent the unity of the male and female divine principle and the union and oneness of the being and psyche, as also the two hemispheres of the brain, intuition and intellect and their synthesis in higher knowledge and wisdom. Shiva wearing the garland of skulls and necklace of snakes, ashes on his forehead and body, represent detachment.

It symbolizes overcoming of fear–including fear of death and attachments of worldly existence, falsehoods, obsessions, delusions, and conquering desires–and transforming sexual energy to creative expressions, not fearing desires but able to work with them making them a "thalic" energy, like in botany, not necessarily "phallic" in Freudian terms, where ojas are transmuted into rajas and tejas–purifed energy and light. Thereby developing the ability and the power like the seed to become the tree itself, for the flower to blossom, and yet have the detachment of the lotus flower to dedicate itself into fragrance and beauty. The world will give you pain, but you have to offer your song in return.

Shiva mantra (as Science of sacred words) – known as bija mantra, as seed mantra, its seed syllables/elemental sounds invoke the power from blessing and protection to turn destiny, to obviate, alleviate negative karma, to empower with abundance, prosperity, and to attract health, love and fulfilment, and impart balance and harmony and healing into pranic energy. Since Shiva is trayambakkam–the three-eyed lord and seer of all the universes and galaxies–earth, prithvi, beneath the earth-patala, and heaven–svarga, his mantra has power to have victory over death and bring liberation and immortality (amritam) from the bondage, temporality and limitation of the earthly life. The power of OM-AUM–the Divine Word and Sound in speech and music are pivotal in understanding and experiencing divinity and psycho-spiritual rebirth and regeneration.

Shiva's trident–trisula– represents the multiple dimensions of his personality and the many facets of his magical powers. Also represents the triple function of Shiva – creation, protection, destruction and regeneration related to the microcosm and macrocosm of the five elements of nature–air, water, fire, earth and space. It connotes the three gunas–the building blocks of energy and action which represent the principle of light, movement/propulsion and inertia.

In Shiva as Nataraja, all of Hinduism's principles can be found in Shiva/Brahman/Purusha who are represented as Nataraja, who is danced as Bharat Natyam, who shines as Surya – Sun, and Chandra – moon, who plays as man and woman, appear as the manifest awakening to its transcendental nature. Each layer is truly a representation of the whole, as it unfolds infinity, cascading into multivariegated patterns of paisley – full of colour and vibrancy – that India is famous for. This inspired Benoit B. Mandebrot to recognize in the mathematics of forms of nature – irregular, asymmetric patterns which he named fractals (1983) which have assumed great significance in our understanding and knowledge. This symmetry in asymmetry, a paradox, is known in Indian and Japanese ideas of beauty – colour and design–since very ancient times.

Hindu cosmology as it is replicated in the elements of the Bharat Natyam draws upon the analogy to fractal patterning. The paradoxical nature of this constitutes the nature of the ultimate reality as beyond mind and thought, can only be directly experienced–expressed in visible icons, images in rituals of working participation, sacrifice, (making the act sacred) – a way of transforming and transcending, all limits of realization of one's nature and to energize and transform into a rapturous, artful sacred experience, the performance of which is not merely entertainment but something much more profound.

The dance of life and other metaphors and symbols again, are primarily spiritual metaphors of interconnectedness within interconnectedness, all appearances of names and forms. Shiva's dance as archetype of alchemical design – as healer – surrounded by rim of fire – for light to enter into matter to bring guidance and transformation of different levels of Prana from physical, emotional, psychological, and spiritual crisis are used for leading, awakening and uniting back to the source, to the One great and beyond all pervading source, the initiation of universe – realization of the Ultimate Truth beyond the symbols, beyond the mind, fascination of all fascinations.

In this process is reflected the profundity of jnana (knowledge) and bhakti (devotion) – two facets of the same source, expressing apparently very different dispositions. The dancer in her dance, living is devotion, she is Prakriti by the sacrifice of her energy and attention to and for the One (Supreme) Purusha which is actually her own source. She becomes the Purusha and icon for the audience and offers sacrifice of their attention. Metaphorically, she is Shakti bowing to Shiva.

She is both the offering and the blessing – union of Bhakti and Jnana Yoga, and her dance that of Karma yoga (will to work and dynamic, detached action). She represents the splendour and form of performance – reminding the obvious fact that it is temporary and transitional. Its temporality, mortality depicted through chosen symbols such as mudras – gestures, costumes, colours, decorations, ornaments,-bangles, bells, anklets, garlands and flowers. These are used to honour beauty and joy, perfection, love, and to invoke divine memory of the supreme, against the backdrop of the canvass of eternity as a consecration.

Flowers occupy a very special place in this process and experience in the Hindu life–worship, dance, drama, music et. al. "Spiritual beauty squanders eternity on a beat of time" as Sri Aurobindo points out in Savitri, the modern epic, a Legend and a Symbol – How Love Conquers Death. The energy of Shakti is the flower for Purusha/Shiva, Prakriti/Parvati is the flower for Purusha.

Shiva is also referred and known by his thousand and one names – from Pashupati, Pushkar, Neelkantha, Vishvanath, Shiva, Sambhu, Shankar, Pinakin, Mahadeva, Maheshwar, Gangadhar (as embodying Ganges, the beautiful mermaid and goddess of all rivers in his jatta (matted hairlocks) etc. The Ganges as river symbolizes fertility, health, abundance and dynamic life force—the waters of life – nourisher, healer and beauty. Shiva is capable of destroying and transmuting enemies of the physical (pashu) nature, vital (rakshas) nature, mental (asuric), demonic, diabolic, egoistic obstructions in the path.

Shiva as Nataraja, the Lord of the Dance, is therefore a quintessential symbol for the meaning of life itself – and the dance and its rhythms and melodies the universe itself. In Shiva's dance, the body becomes the spirit.

Maha Mrityunjaya Mantra (Victory over Death) from the Vedas and Upanishads is ubiquitous – a mantra of power and prosperity, blessing and liberation. It is stated in Sanskrit, the great, great grandmother of all Indo-Aryan languages.

Spring Equinox

Let us now turn to the spring equinoctial initiatory cycle. Concerning this there is a doctrine which is both wonderful and strange, and which is based on the operations of Mother Nature herself. It should be remembered that this phrase Mother Nature, when used with its esoteric significancy, includes not only the physical shell of the universe surrounding us, of which we know the existence through our imperfect senses of report, but also more particularly includes the vast and indeed frontierless realms of the spaces of space.

This strange and wonderful doctrine sets forth that the great initiatory adventure upon which the lofty initiant enters at the time of the Spring Equinox is a copy, a duplication, a repetitive event, in our own small human sphere of what actually occurs at cosmical time intervals among the gods. The initiations which take place even today with more or less uninterrupted regularity at the time of the Spring Equinox include, not only the passing through trials and an ultimate resurrection from the personal man of the god within and an ascension into the spiritual realms, at least for a time, of the initiant's percipient consciousness, but includes also what it has been customary to call in Occidental literature dealing with this theme the descent of the neophyte-initiant, however grand his spiritual stature may be, into the Underworld, into those very real but to us utterly invisible realms of space which have their being in cosmic reaches still more material than our gross sphere of physical mayavi-substance.

It would be wrong to consider this Underworld as belonging exclusively to what has been called in theosophical literature the Eighth Sphere, otherwise the Planet of Death, although indeed the Eighth Sphere must be visited by the percipient consciousness peregrinating at the time.

We have thus, then, a picture of the initiation of the Spring Equinox as a phase of the general initiatory cycle, this phase consisting of severe and searching trials of spiritual and intellectual and psychical as well as astral tests on the one hand, and on the other hand as likewise comprising a descent into spheres never traversed in the ordinary course of their development by the peregrinating monads of average human beings, once these monads have begun to manifest in the human stage.

This strange and mysterious doctrine as thus briefly outlined sets forth that on this our Earth, on this holy and solemn occasion, there occurs a repetition or a duplication of what at certain intervals takes place among the divinities. Just as at certain times in the progress of cosmic destiny a certain divinity leaves its own luminous realms in order to "descend," or more accurately to transfer a portion of its own divine essence, into the world of men for the purpose of aiding and helping erring mankind, so exactly does the neophyte-initiant descend or transfer his percipient consciousness into the Underworld in order to learn and also to help the denizens of those gloomy spheres. What the gods from their lofty height do in this connection to help us, this do likewise these great men in spheres below our own.

One may well ask oneself, as one ponders deeply over this profound teaching and begins to sense its extraordinary and puzzling paradoxes, why it is that a divinity at any time should "descend," or project a portion of its essence, into our sphere which it had long aeons ago left behind in its evolutionary progress. The explanation lies in other teachings concerning the nature of our cosmical solar system, as viewed from the spiritual standpoint. We learn that even the gods themselves are under the sway of almighty destiny, that even they in their own lofty spheres make and unmake karma, and begin, and end after bringing to their completion, works of far-reaching influence on the cosmical spaces, and that a certain portion of these divine activities must of necessity reach to and influence most deeply the spheres of men.

When the student of esotericism understands the profound philosophical import of the teachings concerning the real meaning of the triad of Hindu deities called Brahma, Vishnu, and Shiva, he will come to understand why these amazing events just spoken of take place. As Brahma is the evolver and the producer, and as Vishnu is the sustainer and upholder, so is Shiva, who is the particular patron of esotericists, the regenerator because of being the resolver.

To look upon this triad of divinities in the solar system in the manner in which the Hindu exoteric literary works set them forth is entirely to lose the real meaning and the reach of the esoteric teaching concerning them. The three are three individuals indeed, and yet they are one, very much as evolution and involution are twain and yet are essentially one, because nothing can evolve what is within itself before that within has been involved into it. Thus there can be no Brahma or evolver or producer unless the regenerator or resolver in a past cosmic period had already involved the seeds of the universe later to be evolved or produced. Nor could there be any manvantara or sustained course of cosmic life and evolution unless due to the incessant and continuous influence of the sustainer, upholder, and preserver.

Now, then, these three spiritual-divine energies in the solar system, which are distinctly three and yet one in essence, verily are the higher triad of the septenary belonging to the ten principles of our solar cosmos, and therefore exist and work in their sublimity in what is to us utter silence and darkness, because of being the higher three of the solar system's septenary of worlds of life-energy-consciousness.

From time to time, governed strictly by the karma of the solar system, there arises an impulse in the bosom of Maha-Vishnu to manifest a portion of itself, this portion being a divinity; and this impulse or super-spiritual urge cannot ever be denied or set aside. This impulse furthermore has a technical name in esoteric teaching. It is called *bija*, meaning "seed" or, more accurately perhaps, *avatara-bija*—the cosmic seed of the Avataras.

The Avataras appear on Earth at intervals when the spiritual energies are running low amongst us, and the forces of matter are surging in turbulent waves ever higher. It is as if there were a spiritual psychomagnetic strain in the structure of the solar system, resulting in a spiritual-electric discharge of a spiritual energy, something like lightning on earth, this discharge being popularly called the "descent" of the Avatara, thus preserving the stability and the equilibrium of things. Just so, then, in our world, is it in the cases of these great men, these sublime neophyte-initiants, who during the course of their initiation "descend" into the Underworld for the sake of bringing a spiritual light to the beings enchained in the darkness of those gloomy spheres—spheres which to us seem realms of gloom only because we are higher than they.

So closely is all nature knitted together, so intimately and intricately are the strands of the web of life woven, that all nature must be considered as one vast organism; and when there is a lack of some element-energy in any part of the cosmic body, there is an impulse or urge from other parts, possessing this lacking element-energy in abundance, towards the place where such lack exists, and a consequent passage or peregrination or transference of the lacking element-energy to its destination in order that stability and equilibrium of the cosmical structure may be reestablished or maintained.

The initiation periods do not take place by hap or by hazard, nor are they governed by the mere wish or will of human beings, however grand, however sublime, but take place strictly according to the working of the spiritual cosmic magnetisms of the universe. In consequence, the grand neophyte-initiants enter upon their trials and make their journeys into the Underworld because for the time being they are become utterly obedient servants of the law of the universe, and therefore can hardly do otherwise.

From what has been said, therefore, it becomes immediately obvious how grandly does nature's heart beat throughout with compassionate pulse; for what men in their feebleness of speech describe by such phrases as the reestablishing of disturbed equilibrium, or the maintenance of cosmic stability, is but a poor way of expressing the fact of the automatic operation of the cosmic life in restoring the cosmic harmonies, in the readjustment of the cosmic energies, all under the governance and control of the ineffably grand heart of life-consciousness which beats unceasingly and without pause or surcease to the very end of the solar manvantara.

Hence it is that the Spring Equinox in particular and the Avataras are associated in both human thought and in cosmic actuality. Bear in mind that three are the general cases or general instances in which there occur descents or avataric manifestations of spiritual energies into human existence as extraordinarily powerful motors. One is that of the Avataras produced by influence of the *bija* in Maha-Vishnu; the second is the case of the Buddhas; and the third is that occurring at rare intervals among human beings who are neither Avataras nor Buddhas.

Mark well that the Avatara is the descent of the influence, or of a portion, of a divinity through a loaned intermediate bodhisattvic psychological apparatus in order to manifest in human life in a pure human body. The Buddhas incarnate their own spiritual-divine influences, in each case emanating from the Buddha's own inner god, and do this through the entire term of their work in the world of men; and they manifest these spiritual powers in purposes and works of indescribably lofty benevolence and far-reaching beneficence.

The cases of the rare humans who, being neither Avataras nor Buddhas, from time to time embody or become the dwelling places of spiritual-divine rays, are those unusual men or women who, because of a line of karma which is singularly free from crippling and embarrassing shackles of personality, are able to transmit a ray from the higher triads of themselves. This ray penetrates into and sets on fire with its holy flame the brain-mind and emotional apparatus of such men and women.

The cases of these unusual human beings can be instanced or demonstrated by men and women whose entire existence seems to show a spiritual and intellectual power exceeding by far that of the average run of men, and yet they are human beings only. They may, for instance, be great and noble-minded poets with the seer's vision; they may again be great and noble-minded artists, philosophers, humanitarians, or statesmen; but they are men and men only. They are neither Avataras on the one hand, nor Buddhas on the other hand, and their existence is so well known in the various world religions that they have been called by various names, such as saints or holy men, or by other similar titles.

Although these three classes manifesting spiritual-divine rays—so different as among themselves—are the three instances in which the divine-spiritual manifests in the human sphere, it must be particularly noticed that the originating impulse or urge in all these three classes takes its rise in the mysterious *bija* existing and working from dawn to twilight of a cosmic manvantara in the bosom of Maha-Vishnu.

As a last thought in this connection, mark well that there are Avataras also of Maha-Shiva, just as well as there are Avataras of Vishnu, the sustainer of the solar universe; and it is these Avataras of Maha-Shiva, the solar regenerator, which produce perhaps the most widespread and world-shaking effects in the sphere of men.

The duty of some of the Avataras, their characteristic or swabhava, is to preserve and sustain all that is spiritual, noble, good, lofty, and holy; whereas the work of other Avataras is to regenerate, to make over anew, to bring forth from the womb of destiny what is waiting to come to birth. Hence it is that the work of the Shiva-influence has often and always stupidly been called destruction. The profound philosophy of the process has not been grasped by either Occidental or Oriental scholars; but it is obvious that there are times, brought about by the whirling wheel of life, when evil in the course of destiny must be overthrown, when structures and works which have outlasted their times must be destroyed from the foundations up, in order that a newer building and a grander and loftier structure, both spiritually and materially speaking, shall be raised.

Difficult indeed is the theme of thought upon which I have embarked, and I feel urged to utter a word of warning that one not leap to points of conclusion on the supposition that one has grasped the full import of the meaning of the wonderful doctrine that I have so briefly outlined. Remember that the entire solar universe is one vast organism, quivering and palpitating with life throughout all its reaches, and that what men call spirit or what men call matter are but two phases or two aspects or two events of the on rushing, or the on sweeping, of the cosmic life-consciousness-substance working out its incomprehensibly sublime destiny.

Thus it is that our entire solar system can be viewed from two aspects: one as a cosmic body of spheres builded of the fabric of the cosmic consciousness; and from the other aspect it can be looked upon as an amazingly and most intricately intertwined web of spheres existing on many planes, but all under the dominance, and existing within the limits, of our cosmic divinity. Hence every atom is quivering with life, and is an embodied consciousness centre, which we call a monad, and the only difference between atom and god, between the hosts of darkness and the hosts of light, is one of evolutionary unfoldment.

Finally, let us try to understand somewhat of the meaning of the experiences, so full of mystery and danger, that certain ones, more evolved than we, are now undergoing.

Summer Solstice

We now celebrate the third of the great spiritual and psychical events of the esoteric year, the initiation cycle centering in the Summer Solstice; we celebrate in teaching and by spiritual and intellectual suggestion the actual events of the initiations which take place at this time elsewhere on the surface of the globe.

It is a most suggestive thought, and one that we should carry with us always—each one of us as his or her most prized ideal—that anyone belonging to the outer ring of the mystic Body can, if he or she so will, some day pass from the outer ring to an inner ring, and from that inner ring to one still nearer the centre; and so on, until finally, if the disciple prevail in the conquest of self and in the enlargement of consciousness, be shall one day reach the centre, and thence by his own will and act be swept into the initiatory life-currents which will bear him on the mystic pilgrimage, on the esoteric round of experience, and return a willing and self-conscious renouncer of what he knows he can get, but which he refuses in order to remain and to help the world as one of the stones in the Guardian Wall surrounding humanity.

You will remember that the mystic year contains four seasonal points, and that these four seasons in their cycle are symbolic of the four chief events of progress of initiation: first, that of the Winter Solstice, which event is also called the Great Birth, when the aspirant brings to birth the god within him and for a time at least becomes temporarily at one therewith in consciousness and in feeling; a birth which indeed is the birth of the inner Buddha born of the spiritual solar splendour, or the birth of the mystic Christos.

Then, second, comes the period or event of esoteric adolescence at the Spring Equinox, when in the full flush of the victory gained at the Winter Solstice, and with the marvellous inner strength and power that come to one who has thus achieved, the aspirant enters upon the greatest temptation, except one, known to human beings, and prevails; and this event may be called the Great Temptation. With this initiation at the time of the Spring Equinox the Avataras are particularly concerned, forming as they do one of the lines of activity—a god-line, in fact—of the Hierarchy of Compassion and Splendour, although the Avataras are outside the circle of temptation except insofar as concerns the human portion of them.

Then, third, comes the event of the Summer Solstice, at which time the neophyte or aspirant must undergo, and successfully prevail over, the greatest temptation known to man just referred to; and if he so prevail, which means the renouncing of all chance of individual progress for the sake of becoming one of the Saviours of the world, he then takes his position as one of the stones in the Guardian Wall. Thereafter he dedicates his life to the service of the world, without thought of individual progress—it may be for aeons—sacrificing himself spiritually in the service of all that lives. For this reason the initiation at this season of the year has been called the Great Renunciation. Then, finally, comes the fourth and last period of the cycling mystical year, the event of the Autumnal Equinox, which perhaps is the most sublime, but which actually is not as holy as the initiation which we are now commemorating; because in the initiation of the Autumnal Equinox the neophyte or aspirant passes beyond the portals of irrevocable death, and returns among men no more.

One line of this activity, lofty and spiritual but yet not the line of the Hierarchy of Splendour and Compassion, is that followed by the Pratyeka Buddhas. Aeons will pass before these Pratyeka Buddhas reawaken to take up anew the evolutionary journey, the evolutionary pilgrimage.

The Autumnal Equinox is likewise straightly and closely related to the investigation, during the rites and trials of the neophyte, of the many and varied and intricate mysteries connected with death. For these and for other reasons it has been called the Great Passing.

Children of the Sun and Offspring of the Stars: Has it ever occurred to you to ask yourselves why it is that the stars glitter in the violet dome of night; why our sun shines with unceasing glory, pouring forth through aeons after aeons its own substance of light and life and energy; and why, on the other hand, such vast stretches and realms of nature are sunken in apparently cold and crystalline rigidity: asleep, dormant, seemingly unmoving, although indeed pervaded everywhere and throughout (so that not even an atom is deprived of it) by the all-permeant life and consciousness of the Boundless? Have you ever wondered why these two great contrasts exist in the manifested universe—on the one side, light and movement, activity and power, offspring of divinity and of the spiritual energies; and on the other side, relative immobility, rigidity, crystalline somnolence, and the realms of cold and spiritual sleep?

If you have not asked yourselves these questions, you have not yet really awakened; your spiritual souls are not yet stirring consciously within you, and you are asleep, you are dormant.

It is the beasts that ask themselves no such questions as these, as they live within the restricted bounds of their limited consciousness, for it is a consciousness of feeling and of reaction to feeling only, without the divine fire of self-conscious thought, and without that inquisitive intelligence, that thirst for light and knowledge, which characterize man as a son of the Sun and as an offspring of a stellar parent.

Spirit on one side and matter on the other, conscious life on one side and relative immobility and somnolence of consciousness on the other. As we look upon tenfold nature and consider her activities, we realize that we can figurate the situation as a vast army of the sons of light working upon dark and sleeping matter, the sons of light existing in their embodiments between two poles, both of which to our present human consciousness seem to be impenetrable realms of being. What are these two poles? One is the pole of matter, but the other is the pole of spirit which, because of its incomprehensible brilliance and power, is so far beyond all our intellectual conception or loftiest ideation that it seems as impenetrable to understanding as does the nether pole just spoken of, also apparently dark and incomprehensible.

The reason why nature is thus divided in twain to the understanding of us humans is because we observe on one side the hosts of light, and on the other side the hosts of matter; and yet both fundamentally are one, the difference being that the hosts of light are entities more or less progressed towards the pole of spirit, and the hosts of darkness are ruled by the *mamo-chohans*; as, indeed, the light side is ruled by the Hierarchies of Splendour consisting of *dhyani-chohans* in ever increasing ranges of glory, ascending along the ladder of life beyond the reach of our utmost vision, strain we it upwards as we may.

These two, nature's dark side and her light side, are the two eternal pathways, eternal because of being mighty nature herself. We may speak of the upper or light side as being that of the Hierarchies of Compassion, and the lower or dark side as being that of the Hierarchies of Matter; yet both sides are eternally evolving upwards in everlasting progress. After all, these are but two modes of life, for fundamentally the two are one.

As said a great sage and seer of the Far Orient, Lao-tse, when speaking of the Tao:

> *Its upper part is not bright, and its lower part is not dark. Unceasing in action, nevertheless it cannot ever be named, but from action returns again to the spiritual Void. We may call it the form of the formless, the image of the imageless, the fleeting and the indeterminable [and yet it is the ever-enduring]. Go you before it, you cannot see its face; go you behind it, you cannot see its back.*

Without a name by which it may rightfully be called, it is the origin of the spheres celestial and the spheres material. When it has a name men call it the Eternal Mother of all things. Only he who is constantly free from earthly passions can understand its divine essence; but he whose mind is clogged and blinded by passions can see no more than its outer form. Yet these two, the spiritual and the material, though we call them by different names, in their origin are identically one and identically the same. This sameness is a wondrous mystery, the mystery of mysteries. Understanding this mystery is the portal of all initiation.

Children of the Sun, Offspring of the Stars: Are you like the blind unreasoning beast that has no divine curiosity for wisdom and knowledge and love? Or are you becoming like unto the sages and the seers of the ages, who see in all that surrounds them, in every minutest as well as in every greatest thing or event, a key to a cosmic riddle? Think, and pause a moment over the thought. When you consider the glittering orbs above us and our own glorious daystar whom we call Father Sun, has it never occurred to you that these very stars are manifestations of the Hierarchy of Compassion, bringing light and life and love and wisdom into the dark realms of nature's material spheres? Verily it is so!

Every sun that we discern in the midnight sky, every human creature, every dhyani-chohan whose presence we may instinctively feel, is not only an evolving and progressing entity—especially in the cases of the stars and of the gods—but is also an entity which, motivated by celestial love and wisdom divine, each one in accordance with its own karmic powers and to the extent that it may, has halted on its path or advances slowly on its path, in order to give help to the multitudes and hosts of less progressed entities trailing along behind.

Thus a star, our sun for instance, is not only an evolving god in its divine and spiritual and intellectual and psychical and astral aspects, but is also bending towards us from its celestial throne as it were, and thus appears in our own material realms helping us, giving us light, urging us upwards.

These are no merely vain words of an empty poesy, but are suggestive truth. Everywhere around us nature proclaims law, order, regularity, a succession of event ever following event as beings and things are swept along through the ages on the vast bosom of the river of lives; and all this is the work of the Hierarchy of Splendor and Compassion, of which we in our own humble way form on this earth the outmost circle or sphere. It is the same impulse which sways the gods and the Silent Watchers and the starry beings to help those less progressed, that sways the hearts of the Buddhas of Compassion, and of the Masters of Wisdom and Peace and of their chelas, to take the initiation of the Great Renunciation, thus copying in our human realm what takes place in sublime degree among the divinities. An Avatara is but an exceptional case of a peculiar kind, exemplifying the rule of which the Buddhas are the still nobler and outstanding examples of the general case.

Little do men know of the immense love, the divine impulses of compassion, which sway the souls of those who make the Great Renunciation, giving up all hope of personal evolutionary progress, it may be for aeons to come, in order to remain on earth to help their fellows and in the service of the world. Unrecognized, unthanked, ever silent, ever compassionate, ever filled with holy peace, they work steadily on, watching others go past them as the slowly moving river of lives sweeps along in unending flow. There they stand like pillars of light, these great and noble ones. Although they know that some day their reward shall come, a reward beyond all human understanding, nevertheless there they remain through the ages without thinking of their reward, and endure and endure and endure.

Men in the world have no cognizance whatsoever of the mighty hands and powerful wills which hold back certain cosmic forces and elements, lest these forces and elements ravage men because of the ignorant stupidity and blind willfulness of men in invoking, through their selfish emotions and thoughts, cosmic powers of which they actually have no real consciousness. Because these Great Ones are the protecting shields of mankind, therefore are they called the Guardian Wall.

Every man or woman who does a generous, unselfish, and compassionate act is, by the same token and insofar as the compassionate impulse and act extend, a member of the Hierarchy of Compassion and Splendor. Every man or woman who commits an act of selfishness or who follows blindly and solely an impulse of the matter side of himself or of herself is, by the same token and insofar as the impulse and act extend, acting under the influence of the somber and unholy powers of the material world whose chiefs are the dread mamo-chohans presiding at the pralayas.

Every man or woman who does a selfish, evil, or ignoble act, in very truth is taking a stride backwards and is, let us say in passing, by just so much hindering the forward progress of his fellows; for we are all knitted inseparably together into one web of life, into one living organic union.

How beautiful are they upon whose foreheads shines the light eternal, the light of everlasting peace, the light of wisdom, and the illumination of deathless love! They are growing, and growing rapidly, stimulated by the radiant light which pours forth from within the deeps of their own spiritual being. How blessed is their peace, how unspeakably great their happiness, how calm, how majestic, do they appear! What wonderful strength also are they gaining by each such noble thought, by each such noble act!

Men and women who incarnate this spirit of selfless devotion, in however small a degree, are preparing themselves for a future time when they in turn will stand at the door and knock, seeking, asking for, demanding, and demanding with the innate right of embryo gods, this initiation of the Great Renunciation; and then they will find their place as self-conscious workers in the Hierarchy of Compassion and Splendor.

As Lao-tse again says in this connection, when speaking of the Tao, which is at once the cosmic organism in its divine side and the timeless splendor within the aspirant's own breast: "The entire world of men will flock eagerly to him who holds within himself the mighty form and power of Tao. They will come and receive no hurt, but will find rest, peace, tranquillity, and wisdom."

In speaking again of the practical ethics of him who has already made the Great Renunciation and has passed through the holy rites, the great Chinese Master continues:

He that is empty shall be filled; he that is worn out shall be renewed; be who has little shall have all; he who thinks he has much shall go astray. Therefore the sage embraces in thought the cosmic unity, and thereby becomes a model for all under heaven. He is free from self-display, therefore he shines forth; free from self-assertion, therefore he is distinguished; free from self-glorification, therefore he is glorified; free from self-exaltation, therefore he rises above all. Inasmuch as he never strives with others, there is no one in the world who strives with him.

And further, the same sage and seer, in his paradoxes, taught as follows:

> *Therefore the sage, wishing to be above the people, must by his words put himself below the people. Desiring to be ever nobler than the multitude, he must put himself modestly behind them and at their service. In this way, though he has his natural place above them, the people do not feel his weight; though he has his natural place before them, they do not resent it. Therefore all mankind delight to exalt him, and weary of him not.*

The sage expects no recognition for what he does; he achieves merit but does not take it to himself; I have three precious things which I hold fast and prize above all. The first is gentleness; the second is frugality; the third is proper humility, which keeps me from putting myself before others. Be gentle, and then you can be bold. Be frugal, and then you can be most liberal. Avoid putting yourself before others, and you naturally become a leader among men. But in the present day men cast off gentleness, and are all for being bold. They spurn frugality, and retain only extravagance; they discard proper humility, and aim only at being the first. Therefore they shall surely perish.

It must never for a moment be supposed that the Great Renunciation implies an abandonment of any single part of the manifested universe in order that the neophyte or aspirant may devote himself only to following the sole pathway of light. This in itself is a subtly spiritual selfishness which, let men say what they may, is the spirit governing the career of the Pratyeka Buddhas. It is necessary for the neophyte or chela who desires to pass through even the first gateway of the initiation leading to the Great Renunciation to understand that instead of abandoning the world he remains within it, in order, as he grows greater and stronger, wiser and loftier, to serve ever more largely in the cause of all things that are.

The slightest tinge of individual yearning for personal advancement will bar the doors fast against him, for the very core of this initiation is utter self-renunciation. The effort is indeed a titan's labour, for not only must the personal nature be washed clean, but must be utterly transmuted, as far as is compatible with existence in these realms, into becoming a channel or vehicle or mediator between all above the neophyte and all below him and less than he. He must, in consequence, be tested in every fiber of his being before he can even raise his heart to dare the greatest trials which will lead him first into the gloom of the regions of the Underworld—for he must prevail or fail; and later, when his utterly pure heart and indomitable will have carried him safely out of these, he must be tested in loftier spheres, so that no yearning hunger for more light for himself and for communion with the divinities for his own grace can entice him away from his self-chosen path. The path of the Pratyeka Buddha, after all, is a relatively easy one by comparison with the way of the one who has chosen the Great Renunciation; but oh, how inexpressibly beautiful and sublime is the guerdon that comes to the latter in the far distant future when, his work once done, fully accomplished, like the butterfly he frees himself from the chrysalis and, taking wing into the ambient ether where the gods abide, he becomes at one with them, self-consciously a collaborator with them in the cosmic work. But aeons will pass before this stage shall be reached, aeons upon aeons of remaining in our realms of imperfection and often of strife and pain.

But to the one who has made the Great Renunciation there is a joy in the heart which passeth all understanding, the joy of helping and of raising and of leading others up the stairway of life. Power becomes his; faculties hitherto but partly recognized and perhaps unknown develop within him; he becomes cognizant of mysteries of which in the earlier stages of his growth he had but the faintest adumbration, if indeed any intuition of them at all; and the reason is that the farther he advances in his progress the more perfectly, the more completely, the more entirely, does he become a self-conscious mediator of the wisdom and love of the hierarchies above him, who now can work through him as a perfect instrument, willing, self-sacrificing, joyful, strong, and fully capable.

For him no more is there Dead Sea fruit which turns to ashes in the mouth; for him sorrow and pain as men know them have vanished away. He has come to make his own the world's great sorrow and pain; but, marvellous paradox, the unspeakable peace and bliss that are his because he is an utterly unselfish helper transmute the world's sorrow and pain into the greater light and peace of the splendor above and within him. He becomes at one with universal nature, and instinctively works with her in all her labours; and because of this, nature recognizes him as her master and makes obeisance unto him.

There are many grades of those who take the path of the Great Renunciation: there are, first, the loftiest ones, the very gods themselves who lean from their azure thrones, so to speak, and who communicate with those of the same hierarchy but who are less than they. There are innumerable grades still lower down; there are the Buddhas of Compassion; there are the Masters of Wisdom and Peace; there are the high chelas; there are the chelas of lower degree; and there are even ordinary men and women who feel within themselves the upsurging force of the mighty fire of compassionate love which, at times at least, fills their hearts with its flame. Celestial Buddhas, Dhyani-buddhas, Manushya-buddhas, Bodhisattvas, Masters, chelas, inferior chelas, and great and noble men and women—there in brief is the line or ladder of being which forms the Order of Compassion.

As the chela advances into masterhood, as the Master becomes the Bodhisattva, and as the Bodhisattva develops into the Buddha, and so forth, there is a growing self-conscious realization that every individual of this Hierarchy of Compassion and Splendor is the vehicle or mediator of a divine entity which works through him as its human channel; and in the seventh initiation, although nothing more here can be said of that last and greatest of rites, the initiant comes face to face, it may be for a brief instant or even for months or possibly years, with this inspiring and overshadowing divine entity.

It must never be supposed that the Great Renunciation implies that, once taken, this debars one from further initiation. The Great Renunciation implies, rather, that the entity so devoting himself consecrates himself to a series of further and ever loftier initiations, but with the sole and single purpose of rendering himself ever more fit for transmitting the divine light to others less advanced than he, and for that purpose alone.

The Great Renunciation is also an initiation having many degrees, for the Silent Watcher of whatever grade is the first exemplar and outstanding type of one who sits on the threshold of knowledge absolute and of unspeakable peace, and yet enters not but remains before the last and greatest holy of holies in order that those less developed may have a link with the highest.

Every higher grade entered into during the long cycle of initiation before man becomes a Bodhisattva is an awakening within the neophyte of a new plane of consciousness and the consequent coming into lofty personal relation with the different powers and forces and even entities that belong to each plane as they are attained, the one after the other. Initiation is not something which is added unto the growing and expanding consciousness of the neophyte, as brick is added to brick in building a wall; but the steps of initiation represent, each one, a quickening of the evolutionary process.

In other words, initiation in every instance and throughout time is the bringing out or forth into manifested activity of what already exists within the individual. This thought is so important that I must ask you to pause upon it and to ponder it well. You will at once realize that no initiation can possibly take place merely by request or by petition; that therefore it is utterly impossible for anyone successfully to pass through the rites who is not already prepared to do so. It would be impossible—spiritually, intellectually, psychologically, and psychically—to initiate a beast into even the lowest of the initiatory grades, for the simple reason that the respective inner parts of its constitution are not yet functioning together under the direction and control of a self-conscious entity, as is the case with man.

It is upon this great and basic fact of natural fitness that reposes the entire structure of the ethical teaching which the great Masters of the past have given to their disciples. Discipline must precede the Mysteries—not by any Master's mandate, but simply because it is nature's irrefragable law. Man must prove himself to be worthy, and not only worthy but ready, and not only ready but fit, before his knocking at the portal of the sanctum sanctorum can be even heard; and remember that this "knock" is soundless and made without gesture, for it is a movement of the will, intense and determined, combined with an expanding of the consciousness.

How fit would a man be to enter into the dread regions of the Underworld and to face the often dangerous denizens of those realms if be cannot even control his emotional nature or successfully guide the operations of his own will, and if he does not understand the intricate functioning of his own consciousness? Again, how can a man pass safely through the regions of the superior realms of the universe, with what would be to him, in an unprepared state, all their manifold dangers and subtil appeals, if he himself is not already strong in will and expanded in consciousness and therefore fit to enter those realms? It would be as impossible as to ask a beast to take charge of a chemical laboratory or of an electrical works; or, on the other hand, to demand of a beast that it should compose an oratorio or write an outline of a cosmic philosophy that would mightily and persuasively sway the minds of men.

Yet hundreds of thousands, perhaps millions, of human beings today are not far from being ready and fit to undertake the first of the initiatory trials; but so sunken are they in the web and entanglements of material existence, that not only do they not know of these wonderful truths and of the powers lying latent and hid in their nature, but they would not care to attempt the tests even if they knew of the glorious possibilities that are their birthright. Their own ignorance and inertia prevent their advancing; and it is a part of our duty to awaken the minds of our fellow human beings and to open the doors of their hearts to nature's sublime varieties.

I might say in passing that the greatest and simplest preparation for all the various grades of initiation is our daily life. Here one can prove what he is made of; here he can show the stuff that is in him; here he can strengthen his character, evoke his will, enlarge his understanding, expand his heart-life. The Masters judge, or rather test, a beginner, a neophyte taking his first steps, by the way in which he acts in daily life and reacts to the temptations and trials that daily life puts upon him. These remarks, I repeat, are no vain words of an empty theory, but are sheer truth; and you will understand this at once when you remember that life is the great school, and that all the initiations, without a single exception, are but higher grades, the reaching of higher classes, in the school of life—life terrestrial and life cosmic.

Recollect the nature of the constitution of man which is composed of the following fundamentals or bases: first, a divinity derived from a star, the stellar parent for the individual, and each individual has his own. Next, a monodic essence of an intellectual type, called the manasaputra, deriving from the sun. Third, a psycho-emotional apparatus commonly called the human soul or monad, derivative from the moon-chain. And fourth, a psycho-vital-astral apparatus or body derivative from our own globe earth. And over all and within and running through all these is a superdivine, flameless fire of fundamental consciousness which we can generalize by calling it a son of the Boundless, whose habitat is the range of the frontierless spaces of space.

This is man's own individual ladder of life; and he should earnestly and continuously strive with never an instant's intermission to raise his consciousness ever higher along this ladder, from and out of the body to place it in mastery of his psychomental lunar apparatus which he should conquer and control; and thence still higher to become at one with the manasaputric essence living within him; and in future ages to arise out of this into something still more vast and lofty, which is the divine monad with its range of consciousness extending over the universe which we call the Galaxy or Milky Way; and later on, in aeons to come, he shall go higher, and then again higher, and still higher forever.

Thus verily are we born of the moon, children of the sun, offspring of the stars, and inheritors of the cosmic spaces; for space itself is we and we are it, for we and the Boundless are in essence not twain but one.

In these brief remarks I have endeavored to give, by hint and by allusion, some definite and clear ideas of the character and range of the matters comprised under the esoteric term, the initiation of the Great Renunciation. It too has its compensations unspeakably beautiful, and its end is the heart of the universe. Yet why do I say its "end"?

This is but a figure of speech, but a manner of phrasing; for the heart of the universe is indeed boundless Infinitude, and the frontierless deeps of the Divine. Progress, therefore, is endless; the light becomes ever stronger as one progresses along the path; and what the chela would consider the loftiest summits of the Mystic East which he must climb, he finds when he has arrived and has placed his feet upon those distant peaks, that there are immeasurable distances still to go, and of a grandeur and sublimity which even the gods have not attained.

The Shaiva Siddhanta Philosophy

In the books which treat of Shaivism, there is a reference to four schools, *viz.*, the Nakulisapasupata, the Shaiva, the Pratyabhijna and the Rasesvara.

Shaiva Siddhanta is the philosophy of Southern Shaivism. It owes its origin to no single author. It is midway between Shankara's Advaita and Ramanuja's Visishtadvaita. Its literature consists chiefly of: (i) the twenty-eight Saivite Agamas, (ii) the collection of Saivite hymns known as Tirumurai, (*Compiled by Nambi Andar Nambi, the Tirumurai includes the Tirumantiram of Tirumular, the Tevaram of Appar, Sundarar and Sambandhar, and the Tiruvachakam of Manikkavachagar.*) (iii) the collection of the lives of the Saivite saints, known as Periyapuranam, (iv) Meykandar's Shivajnanabodham, (v) Arulnandi's Shiva-jnanasiddhiar, and (vi) the works of Umapati. Tirumular's work 'Tirumantiram' is the foundation upon which the later structure of Shaiva Siddhanta philosophy was built.

The central doctrine of the Shaiva Siddhanta philosophy is that Shiva is the Supreme Reality, and that the Jiva or the individual soul is of the same essence as Shiva, but not identical. Pati (God), Pasu (soul) and Pasa (the bonds), and the thirty-six Tattvas or principles which constitute the world, are all real. The Shaiva Siddhanta system is the distilled essence of the Vedanta. It prevailed in Southern India even before the Christian era. Tirunelveli and Madurai are the centres of the Shaiva Siddhanta school. Even now, Shaivism is a very popular creed in South India. It is a rival school of Vaishnavism.

Characteristics of the Supreme Reality

The Supreme Reality is called Shiva. He is infinite consciousness. He is eternal, changeless, formless, independent, omnipresent, omnipotent, omniscient, one without a second, beginningless, causeless, taintless, self-existent, ever-free, ever-pure and perfect. He is not limited by time. He is infinite bliss and infinite intelligence. He is free from defects, the all-doer and the all-knower.

Lord Shiva is the God of love. His grace is infinite. His love is infinite. He is the saviour and Guru. He is engaged in freeing the souls from the thralldom of matter. He assumes the form of a Guru out of His intense love for mankind. He wishes that all should know Him and attain the blissful Shiva-Pada. He watches the activities of the individual souls and helps them in their onward march. He liberates the individual souls from their fetters or bonds.

The Five Activities of the Lord

The five activities of the Lord (Pancha-Krityas) are: Srishti (creation), Sthiti (preservation), Samhara (destruction), Tirobhava (veiling) and Anugraha (grace). These, separately considered, are the activities of Brahma, Vishnu, Rudra, Maheshvara and Sada Shiva.

Shiva, Shakti and Maya

The Lord Shiva pervades the whole world by His Shakti. He works through His Shakti. Shakti is the conscious energy of Lord Shiva. She is the very body of Lord Shiva. The potter is the first cause for the pot. The stick and the wheel are the instrumental causes. The clay is the material cause of the pot. Similarly, Lord Shiva is the first cause of the world. Shakti is the instrumental cause. Maya is the material cause. Shakti is not the material cause of the universe, because she is of the nature of consciousness (Chaitanya). Shiva is pure consciousness, but matter is pure unconsciousness. Shakti is the intermediate link between the two. Shakti is the reflex of Shiva. It has no independent existence. Shiva assumes this form out of His great love for mankind. Shiva wishes that all should know Him.

Evolution of the Tattvas from Suddha-Maya

The world undergoes evolution for the benefit of the souls. The whole process of creation is for the sake of the salvation of the souls. The world is real and eternal. The world of matter and souls forms the body of the Lord.

The Shaiva Siddhanta analyses the universe into thirty-six Tattvas or principles, as against the twenty-five of the Sankhya. The thirty-six Tattvas arise from Maya, the material cause of the world. Suddha-Maya is Maya in its primal state. From it arise the five pure principles called Shiva Tattva, Shakti Tattva, Sada Shiva Tattva, Isvara Tattva and Suddhavidya Tattva. Shiva functions through these five pure principles.

Maya evolves into the subtle principles and then into the gross. Shiva Tattva is the basis of all consciousness and action. It is undifferentiated (Nishkala Suddha Maya). The Shakti of Shiva starts her activity. Then Shiva becomes the experiencer. Then He is called Sada Shiva, known also by the name Sadakhya, who is not really separate from Shiva. The Suddha Maya becomes active. Then Shiva, the experiencer, becomes the ruler. He is then Isvara, who is not really separate from Sada Shiva. Suddhavidya is the cause of true knowledge.

The Bonds that Bind the Soul

Anava, Karma and Maya

Souls (Pasu) are by nature infinite, all-pervading, eternal and all-knowing like Lord Shiva (Pati). Yet they think that they are finite, limited and little-knowing, ignorant and temporary. This is due to their bonds (Pasa), *viz.*, Anava, Karma and Maya which are called the three Malas or impurities. Anava is the impurity which makes the all-pervading Jiva think itself to be atomic (Anu). It produces the erroneous notion of finiteness. The second impurity or bond is Karma. The soul acts in certain ways on account of its limitation and does good and evil actions. Karma brings about the conjunction of the soul with its body. The results of the Karma have to be worked out in the world. There should be worlds and bodies, in order to experience the fruits of actions and acquire knowledge. These are provided by Maya, the third Mala or bond. Maya is the material cause of the world. The soul gets experience and limited knowledge through Maya.

The soul learns, by long experience, that this Samsara is full of pains and is transitory, and that he can attain eternal bliss and immortality only by attaining Shivatva or the nature of Shiva or God-realisation. He develops Vairagya (dispassion), and Viveka (discrimination between the real and the unreal, the permanent and the impermanent).

Three Orders of Jivas

The Shaiva Siddhantins divine Jivas or Pasus into three orders, *viz.*, Vijnanakalas, Pralayakalas and Sakalas. Vijnanakalas have only the Anava Mala (egoism). Maya and Karma have been resolved. Pralayakalas are those who are free from Maya alone, in the stage of Pralaya. Sakalas have all the Malas (defects), *viz.*, Anava, Karma and Maya.

The Malas affect only the Jivas and not Shiva. Those who are freed from the Malas or impurities attain Shivatva or the nature of Shiva. They are Siddhas or perfected beings.

The Way To The Attainment Of Shivatva Or God-Realisation

You must free yourself from the three bonds, if you want to attain salvation. You must annihilate Maya which is the root of all sins. You must destroy all Karmas which produce rebirth. You must remove the erroneous notion of a finite self.

The three bonds can be removed only through rigorous Tapas, proper discipline, the help of a Guru, and above all, the grace of Lord Shiva. Charya (observance), Kriya (rites) and Yoga (Yama, Niyama, etc.) constitute the discipline. When the aspirant practises in right earnest Charya, Kriya and Yoga, he obtains the grace of Lord Shiva. Then the Lord instructs the soul, reveals Himself and illumines him. Then the soul realises its nature as Shiva (Jnana).

Discipline and grace culminate in Jnana. Jnana is the supreme means of salvation or the attainment of the final beatitude. Karma and other means are only subsidiary to it. They are only auxiliaries.

The attainment of Shivatva or Shiva-nature does not mean complete merging of the soul in Shiva. The liberated soul does not lose its individuality. It continues to exist as soul in God.

Shivatva is the realisation of an identity of essence in spite of difference. The soul attains the nature of Shiva or God, but it is not itself Shiva or God.

The Shakti Yoga Philosophy

In this system of Shakti Yoga philosophy, Shiva is omnipresent, impersonal and inactive. He is pure consciousness. Shakti is dynamic. Shiva and Shakti are related as Prakasha and Vimarsa. Shakti or Vimarsa is the power that is latent in the pure consciousness. Vimarsa gives rise to the world of distinctions. Shiva is Chit, Shakti is Chidrupini. Brahma, Vishnu and Shiva do their functions of creation, preservation and destruction in obedience to Shakti. Shakti is endowed with Ichha (will), Jnana (knowledge) and Kriya (action). Shiva and Shakti are one. Shakti-Tattva and Shiva-Tattva are inseparable. Shiva is always with Shakti.

Shiva-Tattva and Shakti-Tattva

The creative aspect of the Supreme Shiva is called Shiva-Tattva. Shakti-Tattva is the will of Shiva. It is the seed and womb of the entire world.

Shiva has two aspects. In one aspect, He is the supreme, changeless One who is Sachidananda. This is Para Samvit. Nishkala Shiva is Nirguna Shiva. He is not connected with the creative Shakti. In the other aspect, He changes as the world. The cause of the change is Shiva-Tattva. Shakti-Tattva is the first dynamic aspect of Brahman. This Shiva-Tattva and Shakti-Tattva are inseparable.

Shakti—The Ruler of Maya

Maya or Prakriti is within the womb of Shakti. Maya is the matrix of the world. Maya is potential in the state of dissolution. She is dynamic in creation. Maya evolves into several material elements and other physical parts of all sentient creatures, under the direction of Shakti.

There are thirty-six Tattvas or principles in the Shakti philosophy.

Shakti—The Active Aspect of the Immanent God

The power or active aspect of the immanent God is Shakti. Shiva or Brahman is the unchanging consciousness. Shakti is His changing Power which appears as mind and matter. Shakti is the embodiment of power. She runs this world-show. She maintains the sportive play or Lila of the Lord. She is the supporter of the vast universe. She is the supreme Power by which the world is upheld. She is the Universal Mother. She is Durga, Lakshmi, Saraswati, Kali, Chandi, Chamundi, Tripurasundari and Rajarajesvari. She is Lalita, Kundalini and Parvati. There is no difference between God and His Shakti, just as there is no difference between fire and its burning power.

Devi is Shakti of Lord Shiva. She is Jada Shakti and Chit Shakti. Prakriti is Jada Shakti. Suddha Maya is Chit Shakti. Nada, Bindu and the rest are only names for different aspects of Shakti. Shakti is Prakriti, Maya, Mahamaya and Sri Vidya. Shakti is Brahman Itself. Shakti manifested Herself to Lord Shiva in the ten forms as the Dasa-Maha-Vidyas, *viz.*, Kali, Bagalamukhi, Chhinnamasta, Bhuvanesvari, Matangi, Shodasi, Dhumavati, Tripurasundari, Tara and Bhairavi.

Shakti is Chidrupini. She is pure, blissful Consciousness. She is the Mother of Nature. She is Nature Itself. She is Jagat-Janani, Creatrix of the world; Mahishasura-mardini, destroyer of Mahishasura; Bhrantinasini, destroyer of illusion or Avidya; and Daridryanasini, destroyer of poverty.

The world is a manifestation of Shakti. The countless universes are only dust of Divine Mother's holy feet. Her glory is ineffable. Her splendour is indescribable. Her greatness is unfathomable. She showers Her grace on Her sincere devotees. She leads the individual soul from Chakra to Chakra, from plane to plane, and unites him with Lord Shiva in the Sahasrara.

Manifestations of the Divine Mother

The Supreme Lord is represented as Shiva and His power is represented as His consort—Shakti, Durga or Kali. Just as the husband and wife look after the well-being of the family, so also Lord Shiva and His Shakti are engaged in looking after the affairs of this world.

Divine Mother is everywhere triple. She is endowed with the three Gunas, *viz.*, Sattva, Rajas and Tamas. She manifests Herself as Will (Ichha Shakti), Action (Kriya Shakti) and Knowledge (Jnana Shakti). She is Brahma-Shakti (Saraswati) in conjunction with Brahma, Vishnu-Shakti (Lakshmi) in conjunction with Vishnu and Shiva-Shakti (Gauri) in conjunction with Shiva. Hence She is called Tripurasundari.

Radha, Durga, Lakshmi, Saraswati and Savitri are the five primary forms of Prakriti or Devi. Durga destroyed Madhu and Kaitabha through Vishnu. As Mahalakshmi, She destroyed the Asura Mahisha; and as Saraswati, she destroyed Sumbha and Nisumbha with their companions Dhumralochana, Chanda, Munda and Raktabija.

The Abode of the Divine Mother

The abode of Tripurasundari, the Divine Mother, is called Sri-Nagara. This magnificent abode known as Mani-Dvipa, is surrounded by twenty-five ramparts which represent the twenty-five Tattvas. The resplendent Chintamani Palace is in the middle. The Divine Mother sits in the Bindu-Pitha in Sri-Chakra in that wonderful palace. There is a similar abode for Her in the body of man also.

The body is Shakti. The needs of the body are the needs of Shakti. When man enjoys, it is Shakti who enjoys through him. She sees through his eyes, works through his hands and hears through his ears. Body, mind, Prana, egoism, intellect, organs, and all functions are Her manifestations.

The whole world is Her body. Mountains are Her bones. Rivers are Her veins. Ocean is Her bladder. Sun and moon are Her eyes. Wind is Her breath. Agni is Her mouth.

The Indescribable Glory of Devi

The Story of the Yaksha

In the Kenopanishad, it is said that the gods became puffed up with a victory over the Ashuras. They wrongly took the success to be the result of their own valour and powers. The Lord wanted to teach them a lesson. He appeared before them in the form of a Yaksha—a huge form, the beginning and end of which were not visible. The Devas wanted to find out the identity of this form and sent Agni for this purpose. The Yaksha asked Agni: "What is thy name and power?" Agni replied: "I am Agni, Jatavedas. I can burn up the whole universe in a minute." The Yaksha placed before Agni a dry blade of grass and asked him to burn it. Agni was not able to burn it. He ran away from the Yaksha in shame.

The gods then sent Vayu to enquire who he was. Vayu approached the Yaksha. The Yaksha asked Vayu: "Who are you? What is your power?" Vayu replied: "I am the wind-god. I can blow the whole world in a minute." The Yaksha then placed a blade of grass before Vayu and challenged him to blow it away. Vayu could not make it move an inch from its place. He, too, left the place in shame. Last of all came Indra himself. When Indra reached the place, he found that the Yaksha vanished.

Then Uma appeared before Indra and revealed to him the real identity of the Yaksha. She said to Indra: "It is the power of the Divine Mother—and not that of the gods—that crowned the gods with victory. It is the Shakti of Uma or Haimavati, sister of Krishna, that is the source of the strength of all the gods." Shakti is the great Teacher of Jnana. She sheds wisdom on Her devotees.

The Devi Behind the Gods

When Vishnu and Mahadeva destroyed various Ashuras, the power of Devi was behind them. Devi took Brahma, Vishnu and Rudra and gave them the necessary Shakti to proceed with the work of creation, preservation and destruction. She is at the centre of the life of the universe. She is in the Muladhara Chakra in our bodies. She vitalises the body through the Sushumna. She vitalises the universe from the summit of Mount Meru.

The Mother that Protects

Shakti may be termed as that by which we live and have our being in this universe. In this world, all the wants of the child are provided by the mother. The child's growth, development and sustenance are looked after by the mother. Even so, all the necessities of life, life's activities in this world and the energy needed for it, depend upon Shakti or the Universal Mother.

The first syllable which a chiid or a quadruped utters is the name of the beloved mother. Is there any child who does not owe its all to the affection and love of its mother? It is the mother who protects you, consoles you, cheers you and nurses you. She is your friend, philosopher, preceptor and guide throughout your life. The human mother is a manifestation of the Universal Mother. All women are forms of the Divine Mother.

The Scriptures of the Sakta School

The Devi-Sukta of the Rig-Veda, Sri-Sukta, Durga-Sukta, Bhu-Sukta and Nila-Sukta, and the specific Sakta Upanishads such as the Tripurasundari Upanishad, Sitopanishad, Devi Upanishad, Saubhagya Upanishad, Saraswati Upanishad, Bhavanopanishad, Bahvrichopanishad, etc.—all emphatically declare the Mother-aspect of God.

Saktaism—A Universal Cult

He who worships Shakti, that is, God in Mother-form, as the Supreme *Power* which creates, sustains and withdraws the universe, is a Sakta.

Worship of Shakti, or Shaktaism, is one of the oldest and most widespread religions in the world. Everybody in this world wants power and loves to possess power. He is elated by power. He wants to domineer over others through power. War is the outcome of greed for power. Scientists are followers of Shaktaism. He who wishes to develop will-power and a charming personality is a follower of Shaktaism. In reality, every man in this world is a follower of Shaktaism.

Scientists say now that everything is energy only, and that energy is the physical ultimate of all forms of matter. The followers of the Sakta school of philosophy have said the same thing long ago. They further say that this energy is only a limited manifestation of the infinite Supreme Power or Maha Shakti.

Vedanta and Shaktaism

The basis of Shaktaism is the Veda. Shaktaism upholds that the only source and authority (Pramana) regarding transcendental or supersensual matters such as the nature of Brahman, etc., is the Veda. Shakti Vada or Sakta Darsana is a form of monism or Advaita Vada. Shaktaism is only Vedanta. The Saktas have the same spiritual experiences as those of a Vedantin.

Shaktaism speaks of the personal and impersonal aspects of Godhead. Brahman is Nishkala or without Prakriti, and Sakala or with Prakriti. The Vedantins speak of Nirupadhika Brahman (pure Nirguna Brahman without Maya), and Sopadhika Brahman (with Upadhi or Maya) or Saguna Brahman.

It is all the same. Names only are different. It is a play of words or Sabda Jala. People fight on words only and carry on lingual warfare, hair-splitting, logical chopping and intellectual gymnastics. In reality, the essence is One. Clay only is truth; all modifications such as pot, etc., are in name only. In Nirguna Brahman, Shakti is potential; whereas, in Saguna Brahman, Shakti is dynamic.

Shakti-Yoga Sadhana

Shaktaism is not mere theory or philosophy. It prescribes systematic Sadhana of Yoga, regular discipline according to the temperament, capacity and degree of evolution of the Sadhaka. Sadhana means unfolding, rousing up or awakening of the power of Shakti. Shaktaism helps the aspirant to arouse the Kundalini and unite Her with Lord Shiva and to enjoy the supreme bliss or Nirvikalpa Samadhi. A Sakta does Sadhana which helps the union of Shiva and Shakti through the awakening of the forces within the body. He becomes a Siddha in the Sadhana when he is able to awaken Kundalini and pierce the six Chakras. The mode of Sadhana depends upon the tendencies and capacities of the Sadhaka.

Bhava or Attitude

The aspirant thinks that the world is identical with the Divine Mother. He moves about thinking his own form to be the form of the Divine Mother and thus beholds oneness everywhere. He also feels that the Divine Mother is identical with Brahman. The advanced Sadhaka feels: "I am the Devi and the Devi is in Me." He worships himself as Devi instead of adoring any external object. He says: "*Saham*—I am She (Devi)."

The Awakening of Kundalini

The Shakti must be awakened by Dhyana, Bhava, Japa and Mantra Shakti. The Mother, the embodiment of the fifty letters, is present in the various letters in the different Chakras. When the chords of a musical instrument are struck harmoniously, fine music is produced. Even so, when the chords of the letters are struck in their order, the Mother who moves in the six Chakras and who is the very Self of the letters, awakens Herself. The Sadhaka attains Siddhi easily when She is roused. It is difficult to say when and how She shows Herself, and to what Sadhaka. When Kundalini sleeps, man is awake to the world. He has objective consciousness. When She awakes, he sleeps. He loses all consciousness of the world and becomes one with the Lord. In Samadhi, the body is maintained by the nectar which flows from the union of Shiva and Shakti in the Sahasrara.

Pasu Bhava and Divya Bhava

Physical contact with a female is gross Maithuna. This is due to Pasu-Bhava or animal attraction or brutal instinct. Mother Kundalini Shakti unites with Lord Shiva in the Sahasrara during Nirvikalpa Samadhi. This is real Maithuna or blissful union. This is due to Divya-Bhava or divine disposition. You must rise from Pasu-Bhava to Divya-Bhava through Satsanga, service of Guru, renunciation, dispassion, discrimination, Japa and meditation.

Indispensability of Guru's Guidance and Mother's Grace

Shakti Yoga Sadhana is to be practised in a perfect, practical way under the guidance of a Guru who has become perfect.

Guru is indispensable for the practice of Shakti Yoga Sadhana. He initiates the aspirant and transmits the divine Shakti.

No one can free himself from the thralldom of mind and matter without Mother's grace. The fetters of Maya are too hard to break. If you worship Her as the great Mother, you can very easily go beyond Prakriti through Her benign grace and blessings. She will remove all obstacles in the path, lead you safely into the illimitable domain of eternal bliss, and make you absolutely free. When She is pleased and bestows Her blessings on you, then alone you can free yourself from the bondage of this formidable Samsara.

Knowledge of Shakti Leads to Salvation

Knowledge of Shakti leads to salvation. "*Shakti-Jnanam Vina Devi Nirvanam Naiva Jayate*—O Devi! Without the knowledge of Shakti, Mukti cannot be attained"—says Shiva to Devi. The Jiva or the individual soul thinks, when he is under the influence of Maya, that he is the doer and the enjoyer and identifies himself with the body. Through the grace of Shakti and through Sadhana or self-culture, the individual soul frees himself from all fetters and attains spiritual insight and merges himself in the Supreme.

Worship of the Divine Mother, intense faith and perfect devotion and self-surrender, will help you to attain Her grace. Through Her grace alone you can attain Knowledge of the Imperishable. Glory to Sri Tripurasundari, the World-Mother, who is also Rajarajesvari and Lalita-Devi. May Her blessings be upon you all. May you all obtain the grace of Shakti, the Universal Mother and enjoy the supreme bliss of final emancipation.

The Iconographic Genesis of Shiva

The Shiva Linga-Images of Cosmic Manhood in Art and Mythology.

4

Shiva—The Sensuous Yogi

The Iconographic Genesis of Shiva

Shiva, the Mahadeva, represents one of the three visible forms, or the functional aspects of God, namely, the creation, preservation and dissolution, that is, bringing the cosmos into existence, sustaining it and finally withdrawing it from existing. Lord Shiva represents the last of these three aspects, that is, dissolution or destruction of the cosmos.

The other two aspects, the creation and the preservation, are represented respectively by Prajapati or Brahma, and Vishnu. Prajapati Brahma and Vishnu are Vedic gods. In the Rigveda, Prajapati and Brahma are mentioned as two gods, though both almost alike responsible for the act of Creation. Hence, in later Vedic literature, they merge into one entity, and are sometimes alluded to as Prajapati Brahma and sometimes as two synonymous terms alternating each other. In Puranic literature, Brahma gets pre-eminence and the term Prajapati is used only as the other name of Brahma to avoid monotonous repetition of the same nomenclature. Initially, that is, in the Rigveda, Vishnu is a subordinate type of god, but later by Puranic era, he attains the status of the Lord of the universe and the principal Vedic god.

Shiva as such, or as Mahadeva, is not alluded to in proper Vedas. The Rigveda, however, frequently mentions a brown complexioned sun-like brilliant and gold-like glowing animal-skin-wearing entity by the name of Rudra, or Ishan, who, as per the Rigvedic description, is synonymous of a violent non-

Aryan jungle or tribal god capable of subduing, by his mighty arrows, even the most wild of animals. He did not hesitate even to kill human beings and sought delight in such destruction. Hence, the Rigveda is somewhat critical of his wildness and invokes him for not destroying his devotees, their ancestors, offspring, relatives and horses.

It is only gradually and somewhat in simultaneity that the Rigveda softens and sophisticates him into a civil god of Aryan kind and includes him into the Vedic pantheon. The later Vedic literature identifies in Rudra the proto form of the subsequent Shiva. When Puranas perceived the formless God manifest in His triple function, which He performed as the Creator, Sustainer and Destroyer, both initially and finally, as well as always, they chose Shiva to represent one of these functional aspects of Him and elevated him to the status of the Great Trinity.

Shiva, Brahma and Vishnu, Only the Time-Bound Manifestations of the Timeless God

Shiva, as well as Brahma and Vishnu, do not represent God but only His functional aspects, which manifest in Creation, in sustaining the Creation and, finally, in withdrawing the Creation, which occurs after every kalpa, which is the scheduled age of each Creation. Obviously, after the Creation is withdrawn and the kalpa comes to an end, God's functional aspects too disappear and so does the Great Trinity representing them. Thus, the Trinity, with each of Shiva, Brahma and Vishnu having a scheduled life-span, is the time-bound manifestation of the timeless One, that is, the Trinity disappears after its allotted life-span to reappear when the next kalpa begins, but the Omnipresent God neither appears nor disappears because He is always there before the time began and after its scale has exhausted.

In Indian cosmological tabulation, Shiva's life-span is double of the Vishnu's and Vishnu's double of the Brahma's. Brahma's life-span comprises of 120 Brahma years, which are equivalent to 300 million, 9 hundred thousand, 17 thousand and 376 years of human calendar.

Shiva Precedes Trinity-Partners

Shiva, thus different from what the Puranas proclaim, is not Brahma's creation. He rather precedes his Trinity counterparts, Brahma and Vishnu, on time scale. This pre-eminence of Shiva over others as much reflects in their related theological chronology and availability of their iconic representations in visual arts. Brahma and Vishnu have their roots in the Vedas, and not before. Shiva has a pre-Vedic origin, as his worship cult seems to have been in vogue amongst the Indus dwellers, even around 3000 B.C.

The excavations of various archaeological sites in the Indus valley reveal two sets of archaeological finds that suggest the prevalence of the cult of worshipping both, his anthropomorphic as well as symbolic representations. This excavated material includes a number of terracotta seals representing a yogi icon and the phallus type baked clay objects, obviously the votive lings, suggestive of some kind of phallus-worship cult of the non-Aryan settlers of the Indus cities. Seated in meditative posture, the stern looking Yogi figure wears a typical head-dress made of buffalo horns and is surrounded by various animal icons, lion, elephant, buffalo-type bull, rhinoceros etc. and the bird forms above.

In some seals, this Yogi figure consists of three heads. That the symbolic phallus icons and the anthropomorphic representations relate to one and the same entity becomes obvious from the iconographic thrust, which defines the Yogi form. One of the most significant cardinals of this Yogi iconography, and perhaps more so than others, is its well erect and emphatically exposed phallus, similar to the Urddh-ling Shiva icons, a cult of Shiva, which dominated Shaivite sculptural art for centuries from around the period of Kushanas.

These finds, datable to the period from 3000 B. C. to 1000 B.C. or even later, show the continuity of such worship cult till much after the Vedic era. This is further affirmed by the Rigveda itself. The Rigveda at least twice talks of the phallus worshipping non-Aryan tribes and vehemently condemns the practice.

Shiva in Later Vedic Cult and in The Mahabharata

The Vedas, in their later cult, admit into Vedic pantheon the jatadhari holy Shiva with all his manifestations, namely the bow and arrows carrying archer Sharva, the all pervading Bhava, the benevolent Shambhu and the animal-skin wearer Krittivasanah, but do not approve his phallus worship. In Brahmanical order, Shvetashvara Upanishad is perhaps the earliest treatise that refers, though not directly, to this aspect of Shiva-worship with some degree of reverence when it calls him the Lord of all yonis, that is, the commander of genital faculties of all living ones.

It is, however, in the Mahabharata that his phallus worship has been directly alluded to. The Mahabharata widely follows the Indus perception of Shiva. The Mahabharata, in tune with the Indus Shiva, perceives him as Trishira, or Chaturmukha, that is, having three heads, or four, as Digvasas, that is, without cloth, as Urddh-ling, that is, with upward erect phallus, and as yogadhyaksha, that is, the Lord of Yoga. The Mahabharata goes a little ahead and conceived him also as five headed, four facing the four directions and fifth looking upwards, that is the guardian of the entire cosmos. It is from this five headed Shiva concept that his Sada Shiva form seems to have evolved, as these five heads also symbolize five powers-para, adi, icchha, jnana and kriya, that is, all that is perishable, all that is timeless, and the desire, knowledge and act, of which the entire creation comprise.

Mahabharata's epithet of Pashupati for Shiva is also an adherence to the Indus iconography of Shiva image. The Mahabharata perceives him further as Shardularupa, Vyalarupa and in many other animal forms and as Vrishvaha, or Vrishvahan.

The Skand Purana numbers his animal heads as seven, two of which, namely that of the goat and the horse, he had given respectively to Brahma and Vishnu.

Thus again the number of heads comes to the same five as perceived in the Mahabharata. In visual arts, this Mahabharata iconic vision of Shiva has been widely followed.

Shiva's Trishira, Chaturmukha, Yogi, Pashupati, Vrishvaha and Urddh-ling images, whatever their medium, the stone, canvas, metals and so on, are quite in vogue in Indian arts. The animal headed Shiva is a rarity. However, in visual arts, which allow greater scope for imagination to operate, such as painting, Shiva has been depicted sometimes with multiple animal heads, although to avoid inclusion of his human face these heads are planted on the form of Hanuman, who is Shiva's incarnation.

Such Hanuman forms have heads of animals that have attained mythical heights, say, the horse-headed god Hayagriva, the boar-headed Varah, the great eagle Garuda, and the jungle monarch lion or Simha. Such five-headed and ten-armed figures not only carry most of Shiva's attributes in these hands but such figures also stand upon the form of Apasamara, one of the most characteristic features of Shiva iconography. This iconographic perception defines, on one hand, Shiva as Pashupati, the lord of animals, and on the other as containing within him the entire animal world.

Shiva's Pre-Aryan Origin

Obviously, Shiva had a pre-Aryan origin but where, when and how he came into being, or say into human perception, is not known. This much is, however, certain that a god like him was the presiding deity of the Indus inhabitants and he was worshipped as both, iconically as well as symbolically, that is, as Pashupati and Mahayogi and as Ling.

This in all certainties seems to the initial form of Shiva. May be, the Indus inhabitants shared their god with West Asian settlers who worshipped a similar god Teshav. Teshav, too, was a bull riding deity like Vrishvaha Shiva. He also carried, like Shiva, a trident, pinakin, the bow, arrows, which shot as lightening, danda, the rod, parashu, the axe, and so on. Incidentally, Teshav's consort was also named Maa and was worshipped as Jaganmata, that is, the world mother. Her name so much corresponds with Shiva's consort Uma who too is worshipped as Jagat-janani, the mother of the world. Jaganmata sounds so much like Indus Mother Goddess. Both, Shiva's consort Uma and Teshav's consort Maa rode a lion. Images of

Jaganmata, recovered in excavations, have honeybees hovering around her face. One of the Uma's forms so closely resembles with this honeybee hovering image of Maa. Markandeya Purana alludes to Uma's relation with honeybees, or bhramaris, when it calls her as Bhramaridevi. May be Shiva's consort had some prior tradition of her association with honeybees. It is for such reasons that the known historian Roy Chowdhari, in his Studies in Indian Antiquities, emphatically holds that Rudra-Shiva had some kind of genetic relationship with various gods whose images have been recovered from Anatolia, Mesopotamia and Indus Valley.

Shiva in Vedic Pantheon

Whatever Shiva's origin, the pre-Aryan or from Brahma's frown, as claims the subsequent Puranic tradition, the all assimilating Aryan culture and Vedic religious cult elevated him into its own Order and placed him always on par with its other two great gods, Vishnu and Brahma, and sometimes even above them.

Later Vedic literature invested him with various attributes and details of his person. He has been conceived as thousand eyed, animal skin clad and as possessed of long hair braided into a crown-like shape, the jatamukuta, blue neck, black abdomen, blood-red back and as containing in him all medicinal herbs and drugs, that is, possessed of the power to redeem every one of all kinds of ailments and the cycle of birth and death.

Thus, Vedas perceived him initially as the violent jungle god of non-Aryan kind but later they discovered the other aspect of his being, that is, the well meaning benevolent Shiva. It was this perception of Shiva that seems to have prevailed all after and defined his all subsequent forms, manifestations and visions. Brahmans and Upanishads identify this Vedic perception as Shiva's two aspects, one that of the destroyer and the other of the auspicious benevolent divinity. The Mahabharata identified these two aspects as Ghora and Shiva. Of these Ghora has been equated with fire and Shiva, also mentioned as Maheshvara, has been vested with the deeply spiritual and auspicious saumyarupa, that is, serene and sublime divine being.

Shiva in Myths and Legends

In the course of time, the tradition of faith, both oral and scriptural, and the folk and urbanized, wove around Shiva hundreds of myths and legends and invested his image and visual forms with numerous new dimensions and meaning. The violent jungle god of Vedas and the grim looking horn wearing Yogi of Indus emerges upon the altar of the believing ones, on painter's canvas, in metal casters' mould and in the strokes of hammer and chisel, as the harmless Bholanath, the innocence Lord and the good incarnate, as the supreme auspice, the most formidable of divine powers, the paramount lover and the holiest model of the Vedic family cult.

The term Shiva becomes synonymous of the 'auspicious', good and well being and in him alone, India's all-time maxim, 'Satyam, Shivam, Sundaram', that is, he alone is truthful, benevolent and beautiful, finds its true meaning. In his context, love becomes a divine phenomenon and family the holiest institution. He never codifies his conduct nor sets it to any established rule, but he is all the way the most devoted husband, who passionately loves his consort, and a unique father. He marries Sati, the daughter of Brahma's son Daksha-Prajapati against her father's wishes. Daksha organizes a great yajna and to slight Shiva does not invite him.

Sati, in hope to rectify her father's error, goes to attend the yajna, though Shiva does not approve it. Instead of correcting himself, Daksha humiliates Sati also for marrying a tribal brute. Sati, unable to bear her husband's insult by her father, ends her life by immolating herself into yajna-fire. The outraged Shiva, who madly loved Sati, longed to avenge Daksha's act and created out of his frowns Virabhadra, a young warrior endowed with all of Shiva's powers to destroy Daksha's yajna.

After Virabhadra has destroyed the yajna, entire yajna-bhumi and the capital of Daksha, Shiva retires to forest and wanders in wilderness for thousands of years till Uma, the daughter of Himalaya, and hence also known as Parvati, that is, one born of the Parvata, or mountain, is able to win his love by her long rigorous penance. This time he has in Uma, or Parvati, not a mere consort he loved madly but also the most accomplished woman possessed of paramount beauty, the most caring and devoted wife and as much loving mother.

To complete the holy family, they have, or have been conceived with, five sons, two, Karttikeya and Ganesha, the real ones, and three, Vanasura, Virabhadra and Nandin, the adopted ones, though none of the five were born of his consort's womb. Ganesha was born of Parvati's body elements and Karttikeya those of Shiva.

Brahma Chastised and Shiva's Repentance

Indian mythology accounts how Brahma, the creator of all beings and all things, was fascinated by the beauty of his own created Saraswati, and thereby his daughter. To escape her father's notice, Saraswati turned herself into a female deer. But Brahma did not fail to take note of it and converted himself into a male deer and began chasing her to have sex with her. The moral being as Shiva was, he did not approve a father molesting his own daughter.

He did not fail to notice this immorality of the deer turned Brahma when he saw him chasing Saraswati disguised as she-deer and to chastise him, he, the great archer as he was, shot at Brahma, the male deer. To save himself from Shiva's arrows Brahma returned to his real form but not before he had incurred some loss. He had lost one of his five heads.

Whatever Brahma's immorality, Shiva's act amounted to Brahma-hatya, the sin of killing a Brahmin. As the related legend has it, the sin of the Brahma-hatya rose from where Brahma's head fell and stuck to his wrists. Failing to free himself of it, Shiva sought advice and was suggested to beg and live on begging as repentance till the Brahma-hatya fell down and freed him from its clutches. With the kapal, the skull made of Brahma's dissected head, in his hand, Shiva moved to the Oak Forest and wandered there for many thousand years. Ultimately, the Brahma-hatya separated from his body and fell down on earth. It was thus that his Mahabhikshuka and Kapalin forms evolved. Another tradition has it differently. Deer turned Saraswati ran to save herself from Brahma and Brahma to save himself from Shiva's arrows hid in the sky amidst planets and yet lie hidden as two stars. Brahma's fifth head was removed, according to this legend, for a different reason. Brahma and Vishnu often claimed their relative priority over the other. Once they set to settle it and decided that whosoever first discovered the end of Shiva's Jyotirling would be acknowledged as his superior by the other.

The Jyotirling descended deep below the earth and rose above into sky and both ends were unfathomable. Brahma proceeded upwards and Vishnu downward but both ends were far from their reach. Brahma, however, connived with a Champaka or Ketaki flower and using it as witness claimed to have reached his end of the Jyotirling. Annoyed by Brahma's falsehood Shiva appeared bursting the Jyotirling and to chastise Brahma for his lie removed Brahma's fifth head by the nail of his thumb.

Shiva, the Bholanath in the Real Sense of the Term

As he was a moral being, so he was simple, innocent, generous, benevolent and easily manageable, and hence, even the wicked ones often won his favour and boons of invincible powers and sometimes used them even against him. He, however, as readily punished them when he knew their designs and intentions. Ganga was mad in love for him and wished to unite with him by whatever mean, fair or fowl. When Bhagiratha did rigorous penance to bring Ganga from heaven to the earth for his ancestors' death rituals and redemption, Ganga designed to fulfil her long cherished desire of reaching Shiva.

She appeared before Bhagiratha and agreed to emerge on the earth but warned at the same time that her current, unless Shiva took her on his head, would cleave the earth. Bhagiratha underwent another round of penance, pleased Shiva and got his prayer granted. But, when Ganga landed on his head and showed her supremacy, Shaiva kept her arrested into his hair till she herself prayed him to let her be released. For long containing Ganga into his hair, Shiva becomes known as

Gangadhara Shiva. It was the same with Jalandara, who was caused by Shiva himself. Shiva had opened his third eye for punishing Indra but on Brihaspati's intervention let the fire emitting from it fall into the ocean.

Out of this fire and from ocean's womb rose a male child. As he rose from jala, the water, he was named Jalandara. Later, when he grew into a gold-like glowing youth, he was married to the daughter of Kalanemi, the founder-father of demon clans. Jalandara was now exceptionally powerful and wished to drive out Indra and his crew from Indraloka. Indra prayed Brahma for help but he was helpless against his might. Vishnu declined to act against him, as, being ocean born, he considered him his brother-in-law.

Finally, the great sage Narad incited Jalandara to obtain Parvati, the most beautiful woman in all three worlds, and thus put him against Shiva, as he knew that Shiva alone could destroy him. Arrogant Jalandara challenged Shiva to hand him over his consort and in the process became victim of Shiva's wrath and got killed. Something of the similar kind happened in the case of Ravana, the king of Lanka. Pleased by his penance Shiva blessed him with the boon of immortality. This bred in Ravana vanity and arrogance. This vain and arrogant Lanka ruler wished to have Mount Kailash, the abode of Shiva, shifted to Lanka. He went to Kailash and to uproot it began shaking it. His act of uprooting it sent tremors across the Mountain.

Shiva perceived Ravana's arrogance and was annoyed. To punish Ravana he pressed the Mountain by the thumb of his foot, but before it crushed Ravana, he prayed for Lord's mercy and the compassionate Lord forgave him. Out of this compassionate nature of Shiva there emerged his Ravananugraha-murti, that is, the form of him who was kind to Ravana.

Shiva in Saumya and Raudra-rupas

Thus, Shiva's divine perception as well as iconic visualization developed into two directions, one growing out of his serene sublime benevolent Saumyarupa and the other out of his awe-striking Raudra-rupa. Even in his Saumyarupa, contrary to his Vaishnava counterparts, that is, Vishnu, Brahma or even Indra, whom Puranas define using feudal terms and iconography, Shiva is a simpler being, an amalgam of both, the Raudra and the Saumya rupas.

In both aspects, jatamukuta is his crown, elephant hide his cloak, lion skin his loincloth, snakes his necklace, yajnopavita and other ornaments, bhang his favoured drink and the shade of a roadside tree his castle. He is delighted in dance and dances for both, to create as well as to destroy, and in lasya as well as in Tandava and his Tandava is the Anand-tandava as it aims at re-creating and setting the cycle of creation-destruction-and recreation in motion.

He assists Devas, the gods, in their exploits and battles against demons but unlike them and always differently and in mightier way. Both, the gods and the demons, wish to be immortalized and for obtaining the immortalizing nectar join hands to churn ocean, which contained such nectar. But before the ocean yields nectar, there emerges from it the all-annihilating venom. Even by its vapours it begins to suffocate the entire creation. All, gods and demons, flee to save their lives leaving the creation to its destiny. Shiva comes to rescue. He deposits the venom into his throat and saves the creation from its devastating effect. Stored perpetually in the throat, the venom renders it blue and gives Shiva yet another name of Neelakantha, that is, the blue throated one.

It was in consideration to such exploits that in subsequent days the Vaishnavites and Shaivites were seen with daggers-drawn on the question of the pre-eminence of their respective gods. Ultimately the wise ones of both sects had to discover for the votive images the Harihara form, which combined Hari and Hara, that is, Vishnu and Shiva, into one sanctum image and inspired sectarian unity.

In his purer Raudra-rupa, besides what the Vedas and Puranas perceived in it, these aspects farther expand. He is now perceived as Bhairava, Kapalika, Kalabhairava, Mahakala and in similar other terrific forms. He is the presiding deity of cremation ground, which is his loving abode. He rejoices dancing around a burning pyre and as much upon a dead body. The dark nights, when howls of jackals, wolves and other ignominious animals echoed, are his chosen hours to operate. These jackals and other animals living on human flesh are, otherwise too, his best companions. Bhairava wears around his neck the garland of human skulls and around his waist the girdle of dismembered human hands. Now, besides snake ornaments, scorpions make his earrings and ghostly spirits dance around him. The human skull is his cup and ashes of a burnt corpse his talc, with which he smears and adorns his body. In ritual worship, wine and flesh are his chosen offerings.

In these terrific forms of Shiva Kali, Smashan-Kali, Mahakali, Chhinnamasta, Chamunda, Vagulamukhi etc. are his female counterparts, perceived in Puranas often as his consorts. Bhairava, howsoever terrific his form, has his softer aspects when seated under a canopy or riding his Nandin he represents such beautiful musical modes as the Raga Bhairava, or Raga Kedara.

The Shiva Linga—Images of Cosmic Manhood in Art and Mythology

The renowned sage Bhrigu once ventured to the abode of Lord Shiva, wanting to consult him on an important spiritual matter. On reaching, he knocked the door of Shiva's residence. There was no reply. He knocked again, this time a little louder. No avail. The knocking turned to pounding and then to a desperate beating. Finally Shiva emerged, taking his own time. He had his wife Parvati on his left arm.

Obviously, the Great Lord had thought it fit to first complete his lovemaking with the goddess, rather than immediately divert his attention to the venerated ascetic. Incensed at the treatment meted out to him, the sage cursed Shiva that henceforth, since he was so fond of making love, he would be worshipped in the image of his organ of generation, rather than his anthropomorphic representation. Thus, to this day, Shiva is worshipped in the form of the male organ of procreation, often alone, and frequently conjoined with the corresponding female organ, which is sculpted as a receptacle to receive Shiva's seed. This representation of Shiva is known as the lingam. The word lingam literally means a 'sign' or distinguishing mark. Thus says the Linga Purana: "The distinctive sign by which one can recognize the nature of something is called lingam."

There are variations on the birth of this symbol of Lord Shiva, some of which ascribe an esoteric and abstract origin to it. For example when Shiva is visualized as the intangible primordial Creative Power, the lingam is said to be his sign (symbol) which can be worshipped by his followers, who require a concrete entity to focus their prayers on. This stream of thought however does not negate the phallic connotations of the lingam. Its literal meaning as a distinguishing mark links these two interpretations. Consider a newborn male child. What is the sign which distinguishes its sex?

Thus, the phallus is the lingam, the symbol of Shiva's manhood, and of which the human organ is just a microcosmic reflection.

Another instructive legend describes why the lingam is believed to be one of the most potent emblems in Hindu ideals. It all started with Brahma and Vishnu, who were arguing over their relative supremacy. Their vain arguments were interrupted by a superluminous glow from a strange and blazing pillar, its shape reminiscent of the linga. Both of them sped towards this indescribable flaming light, which grew before their eyes into infinity, piercing the earth and extending through the heavens. Overwhelmed and terrified by the unfathomable vision, the two gods decided to seek the beginning and end of this burning immensity.

Brahma taking the form of a swan flew upwards, and Vishnu dove down acquiring the shape of a boar. Both of the gods however, could not fathom the extent of this fiery column at either end, and returned exhausted and bewildered to the level they had started from. At that moment, the central part of the pillar split open and Shiva revealed himself in his full glory. Overawed, both Brahma and Vishnu bowed before him. Thunderous laughter, or the sound of AUM, issued from the pillar, filling the sky.

Primarily, the glowing, flaming linga was a pillar of fire, connecting heaven and earth. It had no end and no beginning, but it had direction, upwards, as does the earthly fire. In metaphysical terms, it was (is), the vertical axis which both holds apart and joins heaven and earth, dividing and uniting them at the same time, an apt symbol of cosmic integrity. Like the Tree of Life, it is both the foundation and support that ensures equilibrium between heaven and earth.

In Vedic hymns, Rudra (an epithet for Shiva) is identified with Agni, who in these sacred texts is deified as the carrier of the sacrificial offerings to the gods for whom they are intended. Hence, Agni is the mediator between men and gods, and acts as a metaphysical bridge between the two, just like the cosmic linga. A pertinent observation here is that every creative process is accompanied by the generation of heat. Indeed the sexual act is nothing but the offering of the seed of life into the sacred fire of love. Hence, Agni, the God of Fire, is eminently suited as a metaphoric emblem of the tejas (creative heat) of Shiva, both metaphysically and physically.

In this context it is interesting to note that in temples where the linga is worshipped, there is often a conical pot (Skt. Dharapatra), kept hanging over it. At the bottom of this vessel is a small hole, from which water drips continuously. The idea is to cool the 'fiery' linga. In the shrine, Shiva is eternally in embrace of the goddess. By entering the sacred enclosure we are, in a sense, interrupting his amours. Thus water is poured to stifle the hotheaded god's temper. Shiva is Bhairava (quick-tempered), but he is also Ashutosh (One who calms down quickly). Indeed, a devotee needs to calm his god before asking for favours.

Another description of the origin of the lingam gives a more sensuous portrayal. According to this view, there once lived in a forest, a group of hermits with their wives. At some point of time, Shiva ventured their way, resplendent in his naked glory. The virtuous wives, the very epitomes of chastity, lost all their moral qualms, and went berserk with desire at the sight of his tempting body. In addition to their restraint, they also shed their inhibitions, ornaments, and clothes, and embraced this naked stranger in a wild, uncontrolled frenzy. In one version, the forest sages gave a similar curse to Shiva as Bhrigu above. A second source says that naturally angered by this unabashed display of passion by their better halves, they snatched out Shiva's phallus and threw it on the ground.

In a sacred twist to the tale, it is believed that it broke into twelve pieces when it struck the earth, and at each of the spots a pilgrimage centre sprung up. These twelve sacred sites are today known as the 'Twelve Jyotirlingas,' and are important destinations in the itinerary of a Shaivite pilgrim. The word jyotirlinga itself means a 'linga of light,' thus further cementing the association of Shiva's phallus with the cosmic pillar of light.

Yet another legend says that the severed linga of Shiva proved to be proverbially too hot to handle. Falling to the ground, it did not come to rest, but moved about, burning anything and everything that came in its way. Even the combined efforts of all the gods could not restrain its fierce fieriness. Finally it was Mother Earth, whom the linga penetrated, who managed to quench it inside her womb.

In this regard it must also be stressed here that though Shiva is visualized as an thyphallic deity, his true message is not a licence to licentiousness.

According to Agehananda Bharati, Shiva's erect organ connotes the very opposite in this context. It stands for 'seminal retention,' and represents complete yogic control of the senses. Shiva's linga is always vertical, pointing upwards as the phallus of an accomplished yogi, with the semen (Sanskrit. *virya*) rising up, rather than discharging itself.

Here a parallel is drawn with the uncoiled energy of kundalini, which rises and climbs the length of its path. Indeed the vertical is the direction of the sacred; it is a symbol of ascent, pointing to heaven and transcendent regions. The spermatozoid substance when reabsorbed through sexual abstinence, nourishes the cerebral matter. Rising, according to yogic formula, through the subtle channels flanking the backbone, it renders the intellectual faculties more acute. The Yogi perceives sexual energy as though it were coiled up at the base of the spine, which is why it is called kundalini (coiled) and likened to a sleeping snake. When, by means of mental concentration, it awakens and unwinds its coils, it rises like a column of fire toward the zenith, toward the top of the skull and pierces it to reach the transcendent worlds. Shiva's liberated phallus represents this illuminating power rising heavenward beyond the material world. Thus is the linga likened to a pillar of light, guiding us to true knowledge.

The Two Images of Shiva

Images of Shiva are of two kinds: iconic (anthropomorphic) and aniconic. The former represents Shiva as a human being while the latter envisages an abstract origin for him. In this manner is Shiva different from other deities. The images of all other deities bestow only sensuous enjoyment since they are invariably represented in an anthropomorphic form, appealing solely to the sense organs. But Shiva grants both enjoyment and spiritual release. As an icon, he has the body of man, but in his aniconic form he is visualized as the cosmic pillar. Yet this pillar also evokes his phallus. As an abstract shape, the pillar symbolizes a purely conceptual reality that cannot be sensed in material terms. Visually however, the shape of the cylindrical pillar with a rounded top resembles that of the phallus.

Also when the time came for Shiva to reveal himself to both Brahma and Vishnu, he did so in the form of a linga. Hence the linga is an object of the greatest sanctity, more sacred than any anthropomorphic image of Shiva. Not surprisingly thus, the innermost sanctuary of all Shiva temples is reserved for the linga, while the outer precincts of the sacred architecture may show him in his human form. Indeed, though his iconic images abound, no such image ever occupies the centre of attention in a Shiva temple, this honour being reserved exclusively for his linga.

According to Stella Kramrisch, the linga of Shiva has three levels of signification, these are:

(1) *As a sign of Shiva:* This is evident in the literary meaning of the word 'linga,' and also in the fact that the linga

fell from Shiva's own body. Indeed, God resides in whatever is part of god.

(2) *The Linga as Phallus:* This is depicted in the tale of the curse of sage Bhrigu, and Shiva's violation of the chaste wives of the ascetics in the forest.

(3) *The Linga as Made up of Xosmic Substance:* Established in the tale of the rivalry of Brahma and Vishnu.

In a different three-fold division, it is believed that the linga contains within itself all the three divinities making up the Indian trinity of Supreme Godhead, namely Brahma, Vishnu, and Shiva. Brahma abides in the lower part that is hidden inside the earth. Vishnu occupies the middle portion of the linga that is covered by the pedestal, and finally there is Shiva, in the top portion that is visible above the pedestal.

The Shiva portion (Rudra-bhaga) is also known as 'puja-ansha,' or the part of the linga that is to be worshipped. The Vishnu part is identified with Devi ('yonis tu jagad-dhatri Vishnu-svarupini). The Rudra-bhaga is said to be masculine, Vishnu part feminine, and the Brahma part neuter. There is a logical framework behind this identification. Shiva's masculinity is obvious because of his phallic connotations, while Vishnu is often visualized as a woman in ancient mythology. This has its origin in the incident, when to deprive the demons of the nectar of immortality, Vishnu took on the form of a woman, aptly named Mohini (one who bewitches).

Lastly Brahma, as the creator, represents that primordial unmanifest state which precedes all creation. In this archetypal state there is no perceptible duality, and no distinction of positive and negative forces. Only when there is a tendency to create does the first spark of duality appear in this undifferentiated stratum. This duality has the character of complementary poles of attraction, which is eventually manifested in the whole of creation by male and female characteristics. Hence Brahma, by virtue of preceding the duality inherent in creation, is non-dual, neither male, nor female.

Most commonly, in the sanctuaries where it is worshipped, the lingam is represented surrounded by the female organ of generation, the yoni. The yoni grasps the lingam, and indeed it is only when the phallus, the giver of semen, is surrounded by the yoni that procreation can take place. From the relation of linga and yoni, the whole world arises. Everything therefore bears the signature of the linga and the yoni. Each individual linga that enters a womb and procreates is a harbinger of divinity, and engaging in a sacred act.

Brahma has for his cognizance, the lotus. For Vishnu it is the discus; and Indra has his thunderbolt. But we ordinary mortals do not carry any of the distinguishing characteristics of these gods. On the other hand, each of us does have either a linga or a womb. Hence all creatures of this world are but a part of the composite image of the Shiva linga, grasped by the womb of the goddess. Since the linga is shown to encompass the trinity (Brahma, Vishnu and Shiva) and also all creatures of the earth, it is safely logical to say that those ancient venerables who conceived of this awesome symbol were right in deducing that the entire living world, nay the entire universe, is a part of the lingam of Shiva.

Worship of the Linga

The sect of Shiva worshippers known as Lingayats are distinguishable by the miniature linga they wear on their bodies throughout their lives. It is kept in a silver receptacle hung around the neck, and is believed to act both as a protective talisman and an amulet to defeat negative influences.

The Lingayata's are a unique community who do not believe in the caste system, and are known for their undiscriminating attitude towards all. Not surprisingly, a Lingayat woman menstruating continues to keep the linga on her body near her heart. This is in contrast to orthodox thought where women traditionally do not associate themselves with sacred imagery during their periods.

The linga is indeed a great equalizer. Any ordinary devotee will testify to this who has seen worshippers, regardless of sex, caste, or creed, washing and pouring generous libations on the linga, while simultaneously caressing it intimately. Also, the linga is always installed at the ground level, while other anthropomorphic deities remain established at a height, beyond the reach of the ordinary worshipper.

The linga is not just the organ of generation, but a sign of the progenitor and the essence of cosmic manhood manifested in the microcosm. By worshipping it we are not merely deifying a physical organ, but recognizing a form that is both eternal and universal.

Shiva—The Sensuous Yogi

There once dwelt in a dense forest a group of hermits engaged in the most difficult of austerities. The hermitage had a large number of knowledgeable and mighty sages, but they were for the most part ritualists, more involved in the actual process rather than appreciating the symbolic significance behind the liturgies they performed.

Lord Shiva in his role of an ascetic mendicant once approached this group of recluses to beg for alms. The force of Shiva's tapas or meditations glowed forth form his auric body. Combined with the spectacular flicker in his eyes, it presented him as extraordinarily handsome. This comely young ascetic, his naked body smeared with ashes, exerted a powerful influence upon the womenfolk of the hermitage. The wives and daughters of the sages rushed out to greet the naked yogi. The hermits were utterly shocked at the sight of this unclad monk who drove their well-born wives and mothers to a demented level of desire. The women came with offerings of fruits and flowers. When they approached Shiva the sensuous yogi, they shed all restraint, taking hold of his hands, pleading for his attentions. They shed away their inhibitions, their ornaments, their clothes, and embraced the naked stranger with the skull in his hands.

The saints were left speechless. Their years of solitude and penance and the hard monastic life were all repudiated by the inexplicable aberrations of their noble wives. Confused, pained, bewildered and also very angry, the sages asked the stranger for his name and identity. Shiva greeted their queries with a silence. Driven to a level of frenzy the same as their chaste women, these sages in their uncontrolled outrage tore off Shiva's organ of generation from his body. But Shiva, the first amongst yogis, remained supremely unaffected both by the women's adoration and the sages' anger.

As soon as Shiva's organ fell to the ground it assumed a gigantic proportion, making everyone aware of the divine status of this handsome ascetic. Thus is said to have originated the emblematic worship of Shiva's organ, popularly known as the Shiva linga.

The rapture of love, the moment of euphoria in which we forget everything else (reason, wisdom, prudence, social rules, human interests etc), is but an image of the mystical bliss. The lover ceases to be himself and becomes one with the object of his/her desire. Indeed, for an instant, he/she ceases to exist as an individual, merging with the other being in totality. The sole reality at that defining moment is the voluptuousness of desire that unites them: "*Just as in the embrace of his beloved, a man forgets the entire world, all that exists within himself and without, so in union with the Being of knowledge, he no longer knows anything, either within or without*" (Brihadaranyaka Upanishad, 4.3.21).

For an instant, one achieves one's true goal, forgets one's own interests, ambitions, problems, and duties, and participates in that feeling of bliss that is one's true and immortal nature. Mystical rapture is a marvellous feeling of pleasure, similar to the effect produced by bhang, the Indian hemp and favourite drink of Shiva.

In order to be genuine, love and rapture of pleasure must be absolutely irrational. They must not be "useful," "normal," or according to law." They must not be a mere procreative act used to beget children for the continuance of our house, to look after us and defend our property. They must not be the outcome of marriage, which stabilizes our social position and represents a communion of interests. True love must be wholly useless and disinterested, far from any idea of family, progeny, or social order. Only then it is pure, true love. This is why the mystical poets sing of illicit love, the love of what does not belong to you (parakiya) and not of what you already possess (svakiya). Loving a wife, or someone who belongs to us, is part of what binds us to the world of forms and not of what can free us from it. According to Alain Danielou, only adulterous, abnormal love can be considered pure and truly free from all ties, and only it can give us some idea of the mystic experience-it is absurd, disinterested, and destructive of all that is human.

Thus we should not wonder at the fact that representations of human love-the search for voluptuous pleasure-recognize none of the limits that social ethics wish to impose.

Hence the conduct of the virtuous ladies in the hermitage though shocking at first sight, is perfectly understandable from the above viewpoint. In fact the story also brings our attention to the fact that these women were more spiritually advanced than their men folk, who were engaged in endless itineraries of rituals whose symbolic significance they were unable to fathom and were thus far away from the true import of these spiritual pratices. The ladies on the other hand were more intuitively fine tuned to appreciate the true nature of physical desire, sprung naturally from their archetypal inner being and in harmony with their primordial nature uncontaminated by man made constructs, including both social and moral.

The canonical iconography of Shiva further shows him with certain characteristic attributes which emphasize his sensuous nature, while retaining his essentially yogic profile. Some of these traits making up the character and personality of Shiva are:

The Dance of Shiva

It is said that man danced before he spoke. He certainly danced before he painted and sculpted reliefs on his walls. All cultures of the world have given dance a ritual status before any formal ritual or liturgy was codified in texts, or recreated through relief or paint.

Yoga, like dance, is much more than a mere physical exercise. It is a holistic way of relating to the body that involves an increasing awareness on all levels: the physical, the mental, and the spiritual. Yoga unites the functions of each of these aspects of our personality. This is true for dance also. Certainly any successful dance performance is characterized by a balanced harmony between the body and spirit. What is suggested here is that dance, like yoga, is a conscious attempt at integrating all the tiers of our existence. It does not negate but on the contrary affirms the sensual nature of our objective physical being, and treats it as fundamental to any attempt at spiritual awareness as our subjective intangible soul.

Dance is thus a spiritual channel, an opening of both metaphysical and sensuous doorways.

Whirling his limbs, gracefully carved as if a woman's, Shiva as Nataraja gyrates to the rhythms of his essentially fleshy dance-an outpouring of sensual stimulation in free and unrestrained exuberance. His dance is both supremely sexual and sublimely spiritual.

He is the god of destruction, his dance too is thus essentially of a similar nature. A ring of flames encircles him.

These are the cremation fires which are ultimately going to consume our mortal bodies. But on the other hand dance is also an act of creation. It brings about a new situation and transforms the perpetrator into a higher realm of reality and personality. Thus the forces gathered and projected in his frantic, ever-enduring gyration are both of creation and annihilation. According to Clarissa Estes, in her book '*Women Who Run With the Wolves*':

> *"To make love. we dance with Death. There will be flowing, there will be draining, there will be live birth and still birth and yet born-again birth of something new. To love is to learn the steps. To make love is to dance the dance".*

Applying the same criterion, we observe that Shiva's dance of death and regeneration is nothing but the recreation of the sexual act itself, which is composed of an interplay of desire, sensuality, highs and lows, and of course an overriding sensation of ecstasy, all an integral part of Shiva's dance.

A poet has beautifully described dance as "nature struggling to express itself, in terms of the joy of the dance." Hence by extension, in the frenzy of the actual physical act of mating can be discerned the ultimate truth of all manifested existence. This truth is that of birth and inevitable death. These are the defining qualities of Shiva's dance, as also of the sexual act, both of which communicate through an exhilarated appreciation of the body, for its own sake.

The Hair of Shiva

Shiva's tresses are long and flowing, and dark as the night is Supra-normal energy, amounting to the power of magic, resides in such a wildness of hair untouched by the scissors. The celebrated strength of Samson, who with naked hands tore asunder the jaws of a lion and shook down the roof of a pagan temple, was similarly said to reside in his uncut hair.

Shiva's hair also supports a crescent moon, a symbol of the female reproductive cycle. Indeed much of womanly charm, the sensual appeal of the Eternal feminine, is also in the fragrance, the flow and luster of beautiful hair. On the other hand, anyone renouncing the generative forces of the vegetable-animal realm, revolting against the procreative principle of life, sex, earth, and nature, to enter upon the spiritual path of absolute asceticism, has first to be shaved.

He must simulate the sterility of an old man whose hairs have fallen and who no longer constitutes a link in the chain of generation. He must coldly sacrifice the foliage of the head. This is most significantly evidenced in the first act carried out by the Buddha when he renounced the royal palace. He severed his long and beautiful hair with his princely blade.

But though the spiritual and even earthly rewards of this ascetic attitude are high, Shiva does not shave or shear his hair, said to be "sweet with many a pleasant scent." Refusing to take advantage of the symbolical and potent devices of self-curtailment and deprivation, the arch-yogi is forever the unshorn male.

Shiva thus accepts the essentially sensual nature of the manifested world. He makes us aware that we can free ourselves from our attachments through the very attachments themselves and not otherwise. According to the Kama Sutra "those that seek liberation achieve it thanks to detachment, which cannot occur except after attachment, since the spirit of humankind is by nature attracted by the objects of the senses."

Nandi the Bull of Shiva

The vehicle of Shiva is a bull (vrishabh or vrisha in Sanskrit). He is the great sprinkler of the seed, and represents the fecundating energy of Kama the God of love.

The bull which wanders about, anxious to find a mate, is taken as the embodiment of the sex impulse. Most living creatures are governed by their instincts; they are ridden over by the bull. They are merely the appendage of their reproductive powers.

But Shiva is the master of lust. He rides on the bull. Only those who are masters of their own impulses can ride on the bull. Thus the image of Shiva atop his bull represents the sexual drive brought under control, though not weakened, through asceticism. As Mahayogi, the god is master of the bull. This is true even when he is with his shakti, and his images therefore often represent him sitting upon its back, poised gracefully and fully in control.

> *"Among those who have mastered the bull you are the bull keeper.*
> *O Lord! Riding on the bull, you protect the worlds."*
>
> *—Lingopasana-rahasaya*

A primary aim of yoga is to transform our mighty sexual potency into spiritual power. Yogis believe that sex energy is the very energy that man can utilize for the conquest of his own self. The sexually powerful man, if he controls himself, can attain any form of power, even conquer the celestial worlds. On the other hand, men of weak temperament are unqualified for great adventures, physical or mental. The sex impulse must therefore never be denied or weakened. Yoga thus opposes exaggerated austerities.

According to Zimmer, noted Indologist, a deity's animal mount is the manifestation of the god's divine essence. Indeed the man of strong powers is the vehicle of Shiva, through whom the deity reveals his own virile nature and powers. The bull of Shiva is hence also called the joyful (Nandi), correspondingly Shiva himself is known as the lord of joy (Nandikeshvara).

Kundalini and the Marriage of Shiva

The metabolic energy called Kundalini is symbolized as Parvati. She is conceived as the serpent power which lies coiled in the lowest chambers of the human body. Kundalini when properly quickened, unfolds her vibrating hoods and by an upward sweep enters the spinal cord and then the brain, and finally unites above the head with Shiva. In mythology, Shiva's wedding with Parvati is the entrance of this serpent power into the Higher Mind which is compared to the snowy mountains of Kailash. Kailash is the symbol of the highest mind and Shiva has his abode on this mountain where silence reigns eternally.

The analogy is between a human wedding which releases the highest ecstasies of the flesh, and the wedding of Kundalini with Shiva, which is a symbol of the highest bliss attainable by an individual soul.

Conclusion

Our body is the instrument of our destiny. Our intellectual mechanism and spiritual being are not independent of the body that shelters and nourishes them. If we wish for success in anything whatever, we must take care of our body: cherish, satisfy, and content it. Yogis condemn abstinence, just as they condemn excess, since both cause imbalance in the physical and intellectual being. A healthy, vigorous, satisfied body, one that is pleasant to inhabit, is the best vehicle and instrument for human and spiritual accomplishment.

Eroticism and pleasure in all its forms are vital for man's intellectual and physical balance. Life is transmitted through the sexual act, and the giving of life is a duty, a debt to be discharged by whoever has received it. Besides its practical utility, however, physical pleasure plays an essential role in our inner development.

It is the image of divine bliss and prepares us and aids us to attain it. A man who strives to be chaste and who fears, condemns, and thwarts physical love can never free himself from the prison of the senses. He weaves around himself a web of obscure frustrations, which will hinder him from realizing his transcendental destiny.

On the other hand, the man who has tasted all kinds of sensual pleasure can gradually turn aside from them, finding greater sensual pleasure in union with the divine. This is no longer renunciation, but liberation. In discovering the divine, the realized man gradually loses interest in earthly things, virtue, honour, vice, and pleasure. He considers the human act of love in the same way that he breathes the perfume of flowers or listens to the song of birds.

Indeed the remark of the saint who said "I have never renounced any vice: it is they who have left me" summarizes the message of Shiva.

In the Puranas, which collect the most ancient mythological and historical legends, Shiva appears as a mysterious and lascivious deity of the primeval forest. He is naked, and his beauty sèduces all beings. The sages practicing austere asceticism are disturbed by the charms of this unconventional god. His virile power is described as limitless. Wandering through the forest, the symbol of the cosmos, always thyphallic, he scatters his seed. From his seed are born plants, metals, and precious stones.

God of eroticism, Shiva is also the master of Yoga, which is described as the method used to sublimate virile power and transform it into mental and intellectual power. He is therefore the "great Yogi." Fittingly therefore, the Kama Sutra designates the various positions adopted in the act of love as asanas, the same term used to describe the postures of Hathayoga.

Although both Shiva and his goddess Shakti are creator deities, the true scope of their union is not procreation, but pleasure and voluptuousness (ananda). A whole world of legend and myth narrates their love. The two opposites, the positive and the negative pole, acquire reality only in their relations with each other. They exist solely in what unites them, in the spark of pleasure that jumps from one to the other. In other words, the immanent cause of the universe, substance, and creation, is voluptuous desire.

The spermatozoid substance placed in the female has a fecundating action, but the same substance, when reabsorbed through sexual abstinence, nourishes the cerebral matter. Rising, according to yogic formula, through the subtle channels flanking the backbone, it renders the intellectual faculties more acute.

The Yogi perceives sexual energy as though it were coiled up at the base of the spine, which is why it is called kundalini (coiled) and likened to a sleeping snake. When, by means of mental concentration, it awakens and unwinds its coils, it rises like a column of fire toward the zenith, toward the top of the skull-the image of the heavenly vault-and pierces it to reach the transcendent worlds inhabited by Shiva. Shiva's liberated phallus represents this illuminating power rising heavenward beyond the material world. Thus is the linga likened to a pillar of light, guiding us to true knowledge.

Lord Shiva's Grace

Lord Shiva (Rudra) is known as the great merciful Lord in Hinduism. His great mantras, the Mahamrityunjaya, which is the mantra of Rishi Vasishta in Rig Veda (VII.59.10), is well-known as the taraka-mantra (mantra of deliverance), as also is his famed Panchakshara mantra (Namah Shivaya), first appearing in the Taittiriya Samhita.

Shiva has many names in the Vedas, as per his forms: Soma (Sundara), Indra (Shiva and Mahakala; Indra means 'Lord of Senses'), Rudra (Bhairava), Agni (Kumara and Rudra). Most often these are all Rudras (forms of Rudra) as are the Maruts, who are his Ganas, Ugra forms and Siddhas, avatars etc. Indra itself not only is cognate to Yogeshvara as 'Lord of Senses', the Yogi who has controlled his sense, but is also synonymous with Ishvara, the later term for God and signifying Shiva. Thus, Mahendra of Vedas is later Maheshvara etc.

Shiva is an extremely loving deity, and, contrary to Western views, his so-called Phallus or the Linga, is actually representing the Vishwastambha or Cosmic Pillar, of which he is. This pillar (also called Skambha) is all over the Vedic texts.

The Vedas show the way in which our forefathers loved and adored Lord Shiva also:

"You, gracious Shatakratu (hundred powered), have always been a Father and Mother to us-so we now pray to you for Bliss." *(Rig Veda.VIII.87.11)*

"You O Lord of the Sense-organs are the Conqueror-you gave effulgence to the (inner) Sun. You are the Cosmic-Worker, the Cosmic-Deity and the Greatest."
(Rig Veda.VIII.87.2)

Unlike the avatars (human incarnations), Shiva is himself the being of consciousness that others seek to attain. As Soma-Sundara, he dwells in Satyaloka or Crown Chakra as Bliss, and as Brihaspati-Akshyobha (Nilakanda) in the Throat as the Divine Word, for example. He is the Cosmic Being (Vaishwanara).

As the Sun (Savitar), Shiva is also addressed as Shiva-Surya and is the Inner-Sun within us, as Svetasvatara states, and asserts it's claim from the fourth chapter of the Taittiriya Samhita to which it belongs, which itself, contains Savitar-Yoga and Sri Rudram.

As Savitar, Shiva can be worshipped with the Gayatri mantra. Gayatri means 'song of deliverance', and thus it's Shakti (power) as the Goddess (Gauri), takes us over all distress, like a ship across the Ocean. This is al idea in the Agni-Durga Sukta of Rig Veda (I.99.1) and elsewhere as the Solar-Fire or Shiva as the Flame of Consciousness (*i.e.*VI.1.5).

Shiva is the great God of Love (Prema). This is not lust, but Divine Love, free from all material qualities. There are many stories in the Brahmanas, Mahabharata and Ramayana, which tell of ancient Aryan peoples attaining the grace of Shiva, many thousands of years before the Christian era. He is thus the oldest form of God on the planet.

Some scholars have seen Shiva as a 'Dravidian God' or an 'Aryan God' etc.-but in Truth, he belongs to all of humanity. He represents the underlying and inherit consciousness of the Cosmos, not limited to culture, planet, name etc., but being the possession of one and all.

Unlike many cultures or religions, as Christianity or Islam that fear God and his wrath-Hinduism's Lord Shiva is impartial and does not hate anybody, and greatly loves even those who have demonic tendencies, if they worship him with love and reverence.

Thus, he bestowed boons upon great Demons like Ravana, Banasura and many others, because of his all-merciful nature. He knows that all beings are merely lead by the three gunas, and thus plays a personal part in their lilas or divine plays, of their self-being, in the physical world.

There have also been many incarnations or avatars of Shiva. Rishi Agastya, Rishi Sanatkumara (later known as Skanda), Adi Shankaracharya, Ashwathama etc. are all avatars of Lord Shiva from the past.

The Great God incarnates or assumes human form, for the purpose of teaching humanity. It is through the avatars-direct decent of Divinity itself that alone can restore the true teachings of the Universe, and guide others to become true Gurus. Krishna, Buddha and Ramakrishna are examples of this, in the Vishnu form of Shiva. To the true devotee, Vishnu, Shakti, Brahma etc. are not different from Shiva, but mere attributes or forms of him, for specific purposes. Indeed, great is the fool who thinks that Krishna and Shiva are different!

Indra is the form of Shiva that rules the Demigods, and Agni the form of Shiva that accepts the sacrifice (as Medhapati, Rudra Bhairava). Brahma is Shiva as the Creator and Shakti is Shiva in the form of the manifest world or the Universe-the world of Maya, and his feminine aspect. Vishnu and his avatars are the form of Shiva as the Cosmic Preserver, ever-incarnating to restore Truth and Law to all beings in the Universe.

Some Rishis are also his incarnations. Rishi Agastya, Rishi Tirumular, Ashwattama, Balarama, Lakshman, Adi Shankara, Dattatreya and others, are all avatars of Lord Shiva, who come as teachers for Humans and other beings.

True Shaivism, is Advaita (non-dualism), as he sees Shiva everywhere and everything in Shiva, in his various forms. Mother Kali is Shiva in the form of the Divine Mother, representing the Nirvana-shakti and Creation (Maya), and even the Robber, Murderer and others are forms of Shiva, as are the Mother, Father and Son-as the Vedic texts also state with various forms of Shiva.

Many Hindus worship Shiva as the Divine Father. Yet, this is not in the same manner as the Western concept-but see Shiva as the personal ever-loving Father, and thus become one with the consciousness in which Jesus saw the Divine Father, personally. Hence they became known as Rudras in the Vedas, 'forms of Shiva', as his own Sons.

There are thousands of Rudras or Maruts in the Universe, who are celestials mind-born from Shiva ad lead by Skanda, who incarnates and also take forms of Siddhas or Rishis, in human-form-but also sometimes display their higher forms, such as Rishi Agastya does.

Why do they take human form? This is in order to teach humanity and other beings, and to enlighten them to join the Shiva-consciousness, and also become sons of Shiva, through various paths, to suit the varied tastes of men, gods and demons.

Shiva in the form of Brihaspati-Akshobhya (aka Neelkantha), is the teacher of the Celestial Beings, and Brihaspati took the form of Shukarcharya (as Maitareya Upanishad states) to teach the demons. As Agastya and Dattatreya, he teaches Humanity also.

Thus, Shiva is Cosmic Love, and does not discriminate between God, Man or Demon, as he knows all are forms of his own Self, wondering about playing a part in the Divine Lila, which the wise call as Maya. All are forms of Shakti, wondering about playing 'hard to get' with their Divine Consort Shiva, of whom they seek to reunite with.

So, how can we attain the grace of Shiva? The answer is, we already have it, but do not now it. It is like a bank account that we have full of money, but do not use it. We must merely 'withdraw the money', meaning chant his mantra or worship him in any manner, with love and devotion, and he shall appear in his graceful form, either in form of Guru, God or teaching.

In this present age of Kali-Yuga, Maya or Cosmic Illusion is at it's best, and hence keeps obscuring Man's true spiritual heritage, like a game of 'hide and seek'. We must merely through aside this Illusion, and reclaim or realize our Spiritual Being.

We should not think we are fallen-as that is the worst type of religion. We are not fallen-we are transcendent free Souls wondering about in the Cosmic Play of Eternity, as noted before, as Shakti trying to hide from Shiva. We just do not know it, as our Karmas and Kali-Yuga have become so powerful, they have deluded us, and we become entrapped in the world of name and form. We see no Shiva-yet he is all around us!

A wise mans sees him everywhere. That is why worship of his many forms as Gods and Goddesses, and their vidyas, are important. If he sees smoke, for example, it is not mere Smoke-it is Space, and it is Shiva in the form as the Shakti, Dhumavati. Likewise, if he sees a Thundercloud, it is not merely a Thundercloud, it is Mahakala or Kali forms of Shiva, and so forth.

Thus, Shiva is a great gift of India to Humanity, as, unlike other cultures, she has not lost this Auspicious Consciousness (Shivachaitanya), and has retained it through her Rishis, to restore it again and again to the races of mankind, throughout the ages. Being the great Cosmic Being of Prosperity, and our own inner being, let us never forget our great Lord Shiva, of whom is our past, present and future, our origin and final refuge, our teacher and most of all, the Cosmic Love.

Kashmir Shaivism

Kashmir Shaivism, with its potent stress on man's recognition of an already existing oneness with Shiva, is the most single-mindedly monistic of the six schools. It arose in the ninth century in Northern India, then a tapestry of small feudal kingdoms. Maharajas patronized the various religions. Buddhism was still strong. Tantric Shaktism flourished toward the Northeast. Shaivism had experienced a renaissance since the sixth century, and the most widespread Hindu God was Shiva.

According to the traditions of Kashmir Shaivism, Lord Shiva originally set forth sixty-four systems, or philosophies, some monistic, some dualistic and some monistic theistic. Eventually these were lost, and Shiva commanded Sage Durvasas to revive the knowledge. Sage Durvasas' "mind-born sons" were assigned to teach the philosophies: Tryambaka (the monistic), Amardaka (the dualistic) and Shrinatha (monistic theistic). Thus, Tryambaka at an unknown time laid a new foundation for Kashmir Shaiva philosophy.

Then, it is said, Lord Shiva Himself felt the need to resolve conflicting interpretations of the Agamas and counter the encroachment of dualism on the ancient monistic doctrines. In the early 800s, Shri Vashugupta was living on Mahadeva Mountain near Srinagar. Tradition states that one night Lord Shiva appeared to him in a dream and told him of the whereabouts of a great scripture carved in rock. Upon awakening, Vashugupta rushed to the spot and found seventy-seven terse sutras etched in stone, which he named the Shiva Sutras. Vashugupta expounded the Sutras to his followers, and gradually the philosophy spread. On this scriptural foundation arose the school known as Kashmir Shaivism, Northern Shaivism, Pratyabhijna Darshana ("recognition school"), or Trikashasana ("Trika system"). Trika, "three," refers to the school's three-fold treatment of the Divine: Shiva, Shakti and soul, as well as to three sets of scriptures and a number of other triads.

Kashmir Shaivite literature is in three broad divisions: Agama Shastra, Spanda Shastra and Pratyabhijna Shastra. Agama Shastra includes works of divine origin: specifically the Shaiva Agama literature, but also including Vashugupta's Shiva Sutras. The Spanda Shastra, or Spanda Karikas (of which only two sutras are left), are both attributed to Vashugupta's disciple Kallata (ca 850-900). These elaborate the principles of the Shiva Sutras. The Pratyabhijna Shastra's principle components are the Shiva Drishti by Vashugupta's disciple, Somananda, and the Pratyabhijna Sutras by Somananda's pupil, Utpaladeva (ca 900-950). Abhinavagupta (ca 950-1000) wrote some forty works, including Tantraloka, "Light on Tantra," a comprehensive text on Agamic Shaiva philosophy and ritual. It was Abhinavagupta whose brilliant and encyclopedic works established Kashmir Shaivism as an important philosophical school.

Kashmir Shaivism provides an extremely rich and detailed understanding of the human psyche, and a clear and distinct path of kundalini-siddha yoga to the goal of Self Realization. In its history the tradition produced numerous siddhas, adepts of remarkable insight and power. It is said that Abhinavagupta, after completing his last work on the Pratyabhijna system, entered the Bhairava cave near Mangam with 1,200 disciples, and he and they were never seen again.

Kashmir Shaivism is intensely monistic. It does not deny the existence of a personal God or of the Gods. But much more emphasis is put upon the personal meditation and reflection of the devotee and his guidance by a guru. Creation of the soul and world is explained as God Shiva's abhasa, "shining forth" of Himself in His dynamic aspect of Shakti, the first impulse, called spanda. As the Self of all, Shiva is immanent and transcendent, and performs through his Shakti the five actions of creation, preservation, destruction, revealing and concealing. The Kashmir Shaivite is not so much concerned with worshipping a personal God as he is with attaining the transcendental state of Shiva consciousness.

An esoteric and contemplative path, Kashmir Shaivism embraces both knowledge and devotion. Sadhana leads to the assimilation of the object (world) in the subject (I) until the Self (Shiva) stands revealed as one with the universe. The goal-liberation-is sustained recognition (pratyabhijna) of one's true Self as nothing but Shiva. There is no merger of soul in God, as they are eternally nondifferent. There are three upayas, stages of attainment of God consciousness. These are not sequential, but do depend upon the evolution of the devotee. The first stage is anavopaya, which corresponds to the usual system of worship, yogic effort and purification through breath control. The second stage is shaktopaya, maintaining a constant awareness of Shiva through discrimination in one's thoughts.

The third stage is shambhavopaya in which one attains instantly to God consciousness simply upon being told by the guru that the essential Self is Shiva. There is a forth stage, anupaya, "no means," which is the mature soul's recognition that there is nothing to be done, reached for or accomplished except to reside in one's own being, which is already of the nature of Shiva. Realization relies upon the satguru, whose grace is the blossoming of all sadhana.

Despite many renowned gurus, geographic isolation in the Kashmir Valley and later Muslim domination kept the following relatively small. Scholars have recently brought the scriptures to light again, republishing surviving texts. The original parampara was represented in recent times by Swami Lakshman Joo. Today various organizations promulgate the esoteric teachings to some extent worldwide. While the number of Kashmir Shaivite formal followers is uncertain, the school remains an important influence in India. Many Kashmir Shaivites have fled the presently war-torn Valley of Kashmir to settle in Jammu, New Delhi and elsewhere in North India. This diaspora of devout Shaivites may serve to spread the teachings into new areas.

Shaivism is not a single, hierarchical system. It is a thousand traditions, great and small. In the broadest sense Shaivism is life itself. Philosophically it may be understood as six major traditions with many similarities and a few differences.

In the search for peace, enlightenment and Liberation, no path is more tolerant, more mysticaL, more widespread or more ancient than Shaivite Hinduism. Through history Shaivism has developed a vast array of lineages and traditions, each with unique philosophic-cultural-linguistic characteristics, as it dominated India prior to 1100 from the Himalayas to Sri Lanka, from the Bay of Bengal to the Arabian Sea. Here we seek to present the essential features of six major traditions identifiable within the ongoing Shaiva con Shaiva Siddhanta, Pashupata Shaivism, Kashmir Shaivism, Vira Shaivism, Shiva Advaita and Siddha Siddhanta.

It should be understood that this formal and somewhat intellectual division, however useful, is by no means a comprehensive description of Shaivism, nor is it the only possible list. In practice, Shaivism is far more rich and varied than these divisions imply. Our discussion of these six schools and their related traditions is based upon historical information. There are wide gaps in the record, but we do know that at each point where the veil of history lifts, the worship of Shiva is there.

The Shaiva Agamas form the foundation and circumference of all the schools of Shaivism. The system of philosophy set forth in the Agamas is common to a remarkable degree among all these schools of thought. These Agamas are theistic, that is, they all identify Shiva as the Supreme Lord, immanent and transcendent, capable of accepting worship as the personal Lord and of being realized through yoga. This above all else is the connecting strand through all the schools.

Philosophically, the Agamic tradition includes the following principle doctrines: (1) the five powers of Shiva: creation, preservation, destruction, revealing and concealing grace; (2) The three categories, Pati, pashu and pasha-God, souls and bonds; (3) the three bonds: anava, karma and maya; (4) the three-fold power of Shiva-iccha, kriya and jnana shakti; (5) the thirty-six tattvas, or categories of existence, from the five elements to God; (6) the need for the satguru and initiation; (7) the power of mantra; (8) the four padas: charya, kriya, yoga and jnana.

As we explore the individual schools and lineages within Shaivism, keep in mind that all adhere to these doctrines. Our discussion necessarily focuses on the differences between one school and another, but this is not meant to obscure the overwhelming similarity of belief and practice among them.

Agamic philosophy and practices are conveyed to the common man through other channels, one of which is the Shaiva Puranas. These oral collections of stories about the Gods are interspersed with Agamic philosophy. A second channel is the Shaivite temple itself, for the construction of the temples and the performance of the rituals are all set forth in the Agamas-in fact it is one of their main subjects.

The priests follow manuals called paddhati, which are summaries of the instructions for worship contained in the

Shaiva Agamas, specifically the shodasha upacharas, or sixteen acts of puja worship, such as offering of food, incense and water. A third channel is the songs and bhajanas of the sants, which in their simplicity carry powerful philosophic import. A fourth is the on-going oral teachings of gurus, swamis, panditas, shastris, priests and elders.

Such matters of agreement belie the fact that Shaivism is not a single, hierarchical system. Rather, it is a thousand traditions, great and small. Some are orthodox and pious, while others are iconoclastic and even-like the Kapalikas and the Aghoris-fiercely ascetic, eccentric or orgiastic. For some, Shiva is the powerful, terrible, awesome destroyer, but for most He is love itself, compassionate and gentle.

For nearly all of the millions of Shiva's devotees, Shaivism is not, therefore, a school or philosophy; it is life itself. To them Shaivism means love of Shiva, and they simply follow the venerable traditions of their family and community. These men and women worship in the temples and mark life's passages by holy sacraments. They go on pilgrimages, perform daily prayers, meditations and yogic disciplines. They sing holy hymns, share Puranic folk narratives and recite scriptural verses. Still, it is useful for us all to understand the formal streams of thought which nurture and sustain our faith. Now, in our brief description of these six schools, we begin with today's most prominent form of Shaivism, Shaiva Siddhanta.

5

Shaivite Schools of Thought

The scriptures of the new religious movements throw some light on the popular religious life of the period. The god Prajapati was widely believed to be the highest god and the creator of the universe, with Indra, known chiefly as Sakra ("the Mighty One"), second to him in importance. The Brahmans were very influential, but opposition had developed to their large-scale animal sacrifices—on both philosophical and economic grounds—and their pretensions to superiority by virtue of their birth were questioned.

The doctrine of transmigration was by then generally accepted, although a group of outright materialists denied the survival of the soul after death. The ancestor cult, part of the Indo-European heritage, was retained almost universally, at least by the higher castes. Popular religious life largely centred around the worship of local fertility divinities (yakṣha), snake-spirits (naga), and other minor spirits in sacred places and groves (chaitya). Although these sacred places were the main centres of popular religious life, there is no evidence of any buildings or images associated with them, and it appears that neither temples nor large icons existed at the time.

Around 500 BC asceticism became widespread, and increasing numbers of intelligent young men "gave up the world" to search for release from transmigration by achieving a state of psychic security. The orthodox Brahmanical teachers reacted to these tendencies by devising the doctrine of the four ashramas (ashramas, "abodes"), which divided the life of the twice-born after initiation into four stages: the brahmachari (celibate religious student); the grihastha (married householder); the vanaprastha (forest dweller); and the sannyasin (wandering ascetic).

This attempt to keep asceticism in check and confine it to men of late middle age was never followed universally, but thereafter Hindu social theory centred on the concept of varnashramadharma, or the duties of the four classes (varna) and the four stages of life (ashramas), which formed the ideal that Hindus were encouraged to follow.

The 3rd century BC was the period of the Mauryan empire, the first great empire of India. Its early rulers were heterodox, and Ashoka (reigned c. 265-238 BC), the third and most famous of the Mauryan rulers, was a professed Buddhist. Although there is no doubt that Ashoka's patronage of Buddhism did much to spread that religion, his inscriptions recognize the Brahmans as worthy of respect. Sentiments in favour of nonviolence (ahimsa) and vegetarianism, much encouraged by the heterodox sects, spread during the Mauryan period and were greatly encouraged by Ashoka. A Brahmanic revival appears to have occurred with the fall of the Mauryas. The orthodox religion itself was undergoing change at this time, however, because of the development of theistic tendencies that centred around the gods Vishnu and Shiva.

Inscriptions, iconographic evidence, and literary references point to the emergence of devotional theism in the 2nd century BC. Several brief votive inscriptions refer to the god Vasudeva, who by this time was widely worshipped in western India. At the end of the 2nd century, Heliodorus, a Greek ambassador from King Antialcidas of Taxila (in Pakistan), erected a large column in honour of Vasudeva at Besnagar in Madhya Pradesh and recorded that he was a Bhagavata, a term specifically used for the devotees of Vishnu. The identification of Vasudeva with the old Vedic god Vishnu and, later, with Vishnu's incarnation, Krishna was quickly accepted.

Near the end of the Mauryan period the first surviving stone images of Hinduism appear. Several large, simply carved figures survive, not representing any of the great gods but rather yakshas, or local chthonic divinities connected with water, fertility, and magic. The original locations of these images are uncertain, but they were probably erected in the open air in sacred enclosures. Temples are not clearly attested in this period by either archaeology or literature. A few fragmentary images thought to be those of Vasudeva and Shiva, the latter in anthropomorphic form and in the form of a lingam, or phallic emblem, are found on coins of the 2nd and 1st centuries BC.

Early Hinduism

The centuries immediately preceding and following the dawn of the Common Era saw the recension of the two great Sanskrit epics, the Ramayana and the Mahabharata (the latter incorporating into it the Bhagavadgita). Although it was the worship of Vishnu, incarnate as Krishna in the Mahabharata and as Rama in the Ramayana, that developed significantly during this period (see below Epics and Puranas), the god Shiva is active in the Mahabharata, and the cult of Shiva developed alongside the cult of Vishnu.

The Rise of the Major Sects

The Vedic god Rudra gained in importance from the end of the Rigvedic period. In the Svetasvatara Upanishad, Rudra is for the first time called Shiva and is described as the creator, preserver, and destroyer of the universe. His followers are called on to worship him with devotion (bhakti). The tendency for the laity to form themselves into religious guilds, or societies—evident in the case of the yaksha cults, Buddhism, and Jainism—promoted the growth of devotional Vaishnavism and Shaivism. These local associations of worshippers appear to have been a principal factor in the spread of the new cults. Theistic ascetics are less in evidence at this time; but a community of Shaivite monks, the Pasupatas, was also in existence by the 2nd or 3rd century AD.

The period between the fall of the Mauryan empire (c. 185 BC) and the rise of the Gupta (c. AD 320) was one of great change, with most of the area of Pakistan and parts of western India being conquered by a succession of invaders. India was opened to influence from the West as never before, not only by its invaders but by way of the sea through the flourishing trade with the Roman Empire. The effects of the new contacts were most obvious in art and architecture.

The oldest freestanding stone temple in the subcontinent has been excavated at Taxila, near Rawalpindi, Pak. During the 1st century BC the Gandhara school of sculpture arose in the same region and made use of Hellenistic and Roman prototypes, mainly in the service of Buddhism. At that time Hindu temples probably were made of wood because no remains of them are extant; however, literary evidence shows that they must have existed.

By the time of the early Gupta empire the new theism had been harmonized with the old Vedic religion, and two of the main branches of Hinduism were fully recognized. The Vaishnavas had the support of the Gupta emperors, who took the title paramabhagavata ("supreme devotee of Vishnu"). Vishnu temples were numerous and the doctrine of Vishnu's avatars (incarnations) was widely accepted. Of the 10 incarnations of later Vaishnavism, however, only two seem to have been much worshipped in the Gupta period. These were Krishna, the hero of the Mahabharata, who also begins to appear in his pastoral aspect as the cowherd and flute player, and the divine boar (Varaha), of whom several impressive images survive from the Gupta period.

The Shaivites were also a growing force in the religious life of India. The sect of Pasupata ascetics, founded by Lakulisa (or Nahulisa), who lived in the 2nd century AD, is attested by inscriptions from the 5th century and is among the earliest of the sectarian religious orders of Hinduism. Representations of the son of Shiva, Skanda (also called Karttikeya, the war god), appeared on Kushan coins as early as AD 100.

Shiva's other son, the elephant-headed Ganesha, patron deity of commercial and literary enterprises, did not appear until the 5th century. Very important in this period was Surya, the sun god, who had temples built in his honour, although in modern times he is little regarded by most Hindus. The solar cult had Vedic roots but later may have expanded under Iranian influence.

Several goddesses began to gain importance in this period. Although goddesses had always been worshipped in local and popular cults, they play comparatively minor roles in Vedic religion. Laksmi or Sri, goddess of fortune and consort of Vishnu, was worshipped before the beginning of the Christian era, and several lesser goddesses are attested from the Gupta period. But the cult of Durga, the consort of Shiva, was only beginning to gain importance in the 4th century, and the large-scale development of Shaktism (devotion to the active, creative principle personified as the Mother Goddess) did not take place until medieval times.

The Development of Temples

The Gupta period (4th-6th century) saw the rapid development of temple architecture. Earlier temples were made of wood, but freestanding stone and brick temples soon appeared in many parts of India. By the 7th century, stone temples, some of considerable dimensions, were found in the Aryanized parts of the country. Originally, the design of the Hindu temple may have borrowed from the Buddhist precedent, for in some of the oldest temples the image was placed in the centre of the shrine, which was surrounded by an ambulatory path resembling the path around the Buddhist stupa (a religious building containing a relic).

Nearly all surviving Gupta temples are comparatively small; they consist of a small cella (central chamber), constructed of thick and solid masonry, with a veranda either at the entrance or on all sides of the building. The earliest Gupta temples, such as the Buddhist temples at Sanchi, have flat roofs; however, the sikhara (spire), typical of the north Indian temple, was developed in this period and with time steadily was made taller. The massive and tall tower of the Buddhist temple of Buddh Gaya, which was in existence in the 7th century, represents the culmination of Gupta temple architecture.

The Buddhists and Jains had made use of artificial caves for religious purposes, and these were adapted by the Hindus. Hindu cave-temples, however, are comparatively rare, and none has been discovered from earlier than the Gupta period. In the

Pallava site of Mahabalipuram, south of Madras, a number of small temples were carved in the 7th century from outcroppings of rock and represent some of the oldest religious buildings in the Tamil country.

The Spread of Hinduism in Southeast Asia and the Pacific

Hinduism and Buddhism had an immense impact on the civilizations of Southeast Asia and contributed greatly to the development of a written tradition in that area. Around the beginning of the Christian era, Indian merchants in comparatively large numbers settled there, bringing Brahmans and Buddhist monks with them. These religious men were patronized by local chiefs, who converted to the new religion. The earliest material evidence of Hinduism in Southeast Asia comes from Borneo, where late 4th-century Sanskrit inscriptions testify to the performance of Vedic sacrifices by Brahmans at the behest of local chiefs. Chinese chronicles attest an Indianized kingdom in Vietnam two centuries earlier. The dominant form of Hinduism exported to Southeast Asia was Shaivism, though some Vaishnavism was also known there. Later, from the 9th century onward, Tantrism, both Hindu and Buddhist, spread throughout the region.

The civilizations of Southeast Asia developed forms of Hinduism and Buddhism that had distinctive local features and were attuned to the local cultures, but the framework of their religious life was essentially Indian. Stories from the Ramayana and the Mahabharata became widely known in Southeast Asia and are still popular there in local versions. The people of Bali (in Indonesia) still follow a form of Hinduism adapted to their own genius. Versions of the Manu-smrti ("Laws of Manu") were taken to Southeast Asia and were translated and adapted to indigenous cultures until they lost most of their original content. Claims of early Hindu contacts farther east are more doubtful. There is little evidence of the influence of Hinduism on China and Japan, except through Buddhism.

Indian Religious Influence in the Mediterranean World

Nearly as dubious as the question of Hindu influence on the religious life of the Far East is its influence on that of the ancient Mediterranean world. The Greek philosopher Pythagoras may have obtained his doctrine of metempsychosis (transmigration, or passage of the soul from one body to another) from India, mediated by Achaemenian (6th-4th century BC) Persia, but similar ideas were known in Egypt and were certainly present in Greece before the time of Pythagoras. The Pythagorean doctrine of a cyclic universe may also be derived from India, but the Indian theory of cosmic cycles is not attested in the 6th century BC.

Nevertheless, it is known that Hindu ascetics occasionally visited Greece. The most striking similarity of Greek and Indian thought is the resemblance between the system of mystical gnosis (esoteric knowledge) described in the Enneads of the Neoplatonic philosopher Plotinus (3rd century AD) and that of the Yoga-sutras attributed to Patanjali, an Indian religious teacher sometimes dated in the 2nd century AD. The Patanjali text is the older, and influence must be suspected, though the problem of mediation remains difficult because Plotinus gives no direct evidence of having known anything about Indian mysticism. Several Greek (*e.g.*, Clement of Alexandria) and Latin writers show considerable knowledge of the externals of Indian religions, but none gives any intimation of understanding their more recondite aspects.

Certain Vaishnava legends, especially those referring to the infant Krishṇa, bearsome resemblance to those of Christianity, and claims have been made by both Hinduism and Christianity that the one influenced the other. There is, however, no definitive evidence for the priority of either one.

The Rise of Devotional Hinduism (4th-11th century)

The medieval period saw the growth of new devotional religious movements centred on hymnodists who taught in the popular languages of the time. The new movements probably began with the appearance of hymns in Tamil associated with two groups of poets, the Nayanars, worshippers of Shiva, and the Alvars, devoted to Vishnu. The oldest of these date from the early 7th century, although passages of devotional character can be found in earlier strata of Tamil literature.

The term bhakti, in the sense of devotion to a personal god, appears in the Bhagavadgita and the Svetasvatara Upanishad. In these early sources it represents a devotion still somewhat restrained and unemotional. The new form of bhakti, associated with singing in the languages of the common people, was highly charged with emotion, and the relation of worshipper and divinity was often described by the analogy of that of lover and beloved.

This devotional poetry is characterized by a mystical fervour not found in the Upanishads and the Bhagavadgita, in both of which, even when the object of meditation is conceived as a personal God, there is little expression of passion. The Tamil "saints," however, felt an intense love (Tamil: anbu) of a personal kind toward their god. They experienced overwhelming joy in his presence and deep sorrow when he did not reveal himself. Some of them felt a profound sense of guilt or inadequacy in the face of the divine. But the dominant emotion in these poems is one of joy, often expressing itself in song and dance. The poems have a strong ethical content and encourage the virtues of love, humility, and brotherhood. The ideas of these poets, spreading northward, probably were the origin of the growth of bhakti in northern India.

The devotional cults further weakened Buddhism, which had long been on the decline. From time to time Hindus, especially Shaivites, took aggressive action against Buddhism. At least two Shaivite kings—the Hephthalite invader Mihirakula (early 6th century) and the Bengal king Sasanka (early 7th century)—are reported to have been active persecutors, destroying monasteries and killing monks. The philosophers Kumarila and Shankara were also strongly opposed to Buddhism. In their journeys throughout India, their biographies claim, they vehemently debated with Buddhists and tried to persuade kings and other influential people to withdraw their support from Buddhist monasteries. Only in Bihar and Bengal, because of the patronage of the Pala dynasty and some lesser kings and chiefs, did Buddhist monasteries continue to flourish. Buddhism in eastern India, however, was well on the way to being reabsorbed into Hinduism when the Muslims invaded the Ganges (Ganga) Valley in the 12th century. The great Buddhist shrine of Buddh Gaya, the site of the Buddha's enlightenment, became a Hindu temple and remained as such until recent times.

At the end of its existence in India, Buddhism developed in a way that had some effect on Hinduism. Among the Buddhist Tantrists appeared a new school of preachers, often known as siddhas (those who have achieved), who sang their verses in the contemporary languages, early Maithili and Bengali. They taught that giving up the world was not necessary for release from transmigration and that by living a life of simplicity in one's own home one could achieve the highest state. This system, known as Sahajayana ("the Vehicle of the Natural," or "the Easy Vehicle"), influenced both Bengali devotional Vaishnavism, which produced sects called Sahajiya with similar doctrines, and the Natha yogis (mentioned below), whose teachings influenced Kabir and other later bhakti teachers.

Hinduism in 11th-19th Century

The Challenge of Islam and Popular Religion

The phase of Indian history marked by the domination of the Muslims in most of northern India saw great changes in Indian religion. The advent of Islam in the Ganges Basin at the end of the 12th century resulted in the withdrawal of royal patronage from Hinduism in much of the area. The attitude of the Muslim rulers toward Hinduism varied. Some, like Firuz Tughluq (ruled 1351-88) and Aurangzeb (ruled 1658-1707), were strongly anti-Hindu and enforced payment of jizya, a poll tax on unbelievers. Others, like the Bengali sultan Husayn Shah 'Ala' ad-Din (reigned 1493-1519) and the great Akbar (reigned 1556-1605), were well-disposed toward their Hindu subjects. Many temples, however, were destroyed by the more fanatical rulers. Conversion to Islam was more common in areas where Buddhism had once been strongest—modern Pakistan, Bangladesh, and Kashmir.

On the eve of the Muslim occupation, Hinduism was by no means sterile in northern India, but its vitality was centred in the southern, Dravidian-speaking areas. Throughout the centuries, the system of class and caste had become more rigid; in each region there was a complex hierarchy of castes strictly forbidden to intermarry and interdine, controlled and regulated by secular powers who acted on the advice of the court Brahmans. The large-scale Vedic sacrifices had practically vanished, but simple domestic Vedic sacrifices continued, and new forms of animal, and sometimes vegetarian, sacrifice had appeared, especially connected with the cult of the Mother Goddess.

By that time, the main divinities of later Hinduism were worshipped. Rama, the hero of the epic poem, had become the eighth avatar of Vishnu, and his cult was growing, although it was not yet as prominent as it later became. Similarly, Rama's monkey helper, Hanuman, now one of the most popular divinities of India and the most ready helper in time of need, was rising in importance. Krishna was worshipped with his adulterous consort, Radha. Strange syncretic gods had appeared, such as Harihara, a combination of Vishnu and Shiva, and Ardhanarisvara, a synthesis of Shiva and his shakti Parvati or Durga.

Temple Complexes

From the Gupta period onward Hindu temples tended to become larger and more prominent, and their architecture developed in distinctive regional styles. In northern India the best remaining Hindu temples are found in the Orissa region and in the town of Khajuraho in northern Madhya Pradesh. The best example of Orissan temple architecture is the Lingaraja temple of Bhubaneswar, built about 1000.

The largest temple of the region, however, is the famous Black Pagoda, the Sun Temple (Surya Deula) of Konarak, built in the mid-13th century. Its tower has long since collapsed, and only the assembly hall remains. The most important Khajuraho temples were built during the 11th century. Individual architectural styles also arose in Gujarat and Rajasthan, but their surviving products are less impressive than those of Orissa and Khajuraho. By the end of the 1st millennium AD the South

Indian style had reached its apogee in the great Rajarajesvara temple of Thanjavur (Tanjore).

In the temple the god was worshipped by the rites of puja (reverencing a sacred being or object) as though the worshippers were serving a great king. In the important temples a large staff of trained officiants waited on the god. He was awakened in the morning along with his goddess, washed, clothed and fed, placed in his shrine to give audience to his subjects, praised and entertained throughout the day, ceremoniously fed, undressed, and put to bed at night. Worshippers sang, burned lamps, waved lights before the divine image, and performed other acts of homage.

The god's dancing girls (devadasis) performed before him at regular intervals, watched by the officiants and lay worshippers, who were his courtiers. These women, either the daughters of devadasis or girls dedicated in childhood, may have also served as prostitutes. The association of dedicated prostitutes with certain Hindu shrines can be traced back to before the Christian era.

It became more widespread in post-Gupta times, especially in South India, and aroused the reprobation of 19th-century Europeans. Through the efforts of Hindu reformers the office of the devadasis was discontinued. The role of devadasis is best understood in the context of the analogy between the temple and the royal court, for the Hindu king also had his dancing girls, who bestowed their favours on his courtiers.

Parallels between the temple and the royal palace also were in evidence in the rathayatras (shrine processions). As on festival days, when the king issued from his palace and paraded around his city, escorted by courtiers, troops, and musicians, so also the god paraded around his city in a splendid procession, together with the lesser gods of the minor shrines. The god rode on a tremendous and ornate moving shrine (ratha), which was often pulled by large bands of devotees. Rathayatras still take place in many cities of India.

The best-known is the annual procession of Jagannatha ("Juggernaut"), a form of Vishnu, at Puri, Orissa. The great temples were (and still are) wealthy institutions. They were supported by the transfer of the taxes levied by kings on specific areas of the nearby countryside, by donations of the pious, and by the fees of worshippers. Their immense wealth was one of the factors that encouraged the Ghaznavid and Ghurid Turks to invade India after the 11th century. They were controlled by self-perpetuating committees—whose membership was usually a hereditary privilege—and by a large staff of priests and temple servants under a high priest who wielded tremendous power and influence.

The great walled temple complexes of South India were (and still are) small cities, containing the central and numerous lesser shrines, bathing tanks, administrative offices, homes of the temple employees, workshops, bazaars, and public buildings of many kinds. Directly and indirectly they played an important part in the economy, as they were among the largest employers and greatest landowners in their areas. They also performed valuable social functions because they served as schools, dispensaries, poorhouses, banks, and concert halls.

The Muslim occupation brought India into close contact with a different, more aggressive, religion. In such circumstances, the absence of a central religious authority in Hinduism was a source of strength. The purohitas, or family priests who performed the domestic rituals and personal sacraments for the lay people, continued to function, as did the thousands of ascetics. In Muslim-occupied territory the temples suffered the most. In the sacred cities of Varanasi (Benares) and Mathura, no large temple remains from any period before the 17th century. The same is true of most of the main religious centres of northern India, but not of the regions where the Muslim hold was less firm, such as Orissa, Rajasthan, and South India.

Sectarian Movements

Before the time the Muslims invaded the subcontinent, the new forms of South Indian bhakti were spreading beyond the bounds of the Dravidian south. Certain Vaishnava theologians of the Pancharatra and Bhagavata schools, including Ramanuja, a Tamil Brahman who was for a time chief priest of the Vaishnava temple of Srirangam, near Tiruchchirappalli (Trichinopoly), taught in the 11th century. They gave the growing Vaishnava bhakti cults a philosophical framework that also influenced some Shaivite schools.

Two other Vaishnava teachers deserve mention. Nimbarka, a Telugu Brahman of the 12th or 13th century, spread the cult of the divine cowherd and his favourite gopi (cowherdess, especially associated with the legends of Krishna's youth), Radha. His sect survives near Mathura but has made little impact elsewhere. More important was Vallabh (Vallabhacharya; 1479-1531), who took the Vaishnava doctrine of grace and emphasized its erotic imagery.

His sect is noteworthy because it stresses absolute obedience to the guru (teacher). Early in its existence it was organized with a hierarchy of senior monks (gosvami), many of whom became very rich. The Vallabhacharya sect was once very influential in the western half of North India, but it declined in the 19th century, in part because of a number of lawsuits against the chief guru, the descendant of Vallabh.

The Shaiva sects also developed from the 10th century onward. In South India there emerged the school of Shaiva-siddhanta, still one of the most significant religious forces in that region, and one that, unlike the school of Shankara, does not admit the full identity of the soul and God. A completely monistic school of Shaivism appeared in Kashmir in the early 9th century. Its doctrines differ from those of Shankara chiefly because it attributes personality to the absolute spirit, who is the god Shiva and not the impersonal brahman.

An important and interesting sect, founded in the 12th century in the Kannada-speaking area of the Deccan, was that of the Lingayats, or Virashaivas ("Heroes of the Shaiva Religion"). Its traditional founder, Basava, taught doctrines and practices of surprising unorthodoxy: he opposed all forms of image worship and accepted only the lingam of Shiva as a sacred symbol. Virasaivism rejected the Vedas, the Brahman priesthood, and all caste distinction. Several Lingayat practices, now largely abandoned, such as the remarriage of widows and the burial of the dead, are deliberately antinomian.

An important development of Shaivism in North India was brought about by Gorakhnath (Goraksanatha), who in the 13th century became leader of a sect of Shaivite ascetics known as Natha ("Lord") from the title of their chief teachers. The Gorakhnathis were particularly important as propagators of the practices of hatha-yoga, a form of yoga that requires complex and difficult physical exercises and that has become popular in the West. These yogis, who are still numerous, influenced the teaching of several of the bhakti poets.

Bhakti Movements

The poets and "saints" of medieval bhakti appeared throughout India. Although all have their individual genius, the bhakti lyricists share a number of common features whatever their language. The Sanskrit education needed for authors of Sanskrit texts limited them largely to the Brahman class and thus put a definite stamp on them. Because bhakti poets could use any language, they might come from any class. They brought to their poetry a familiarity with folk religion unknown or ignored in the Sanskrit texts. The use of the spoken language, even though it was formalized, made possible the immediate expression of an unmediated vision that needed no further context; thus the lyrics are short, intensely personal, and precise. These works illustrate the localistic and reformist tendency evidenced throughout India in the vernacular literatures, especially in Tamil, Bengali, and Hindi.

The origin of the new forms of Hinduism has been attributed to the influence of Islam, but the proposition that the rise of popular emotional bhakti was a response to Islam is impossible, for the practice of singing ecstatic hymns in the current local language was well-known in South India even before Muhammad. All the features of this form of bhakti are found in the Bhagavata-Purana and in the commentaries of Ramanuja.

Ganesha—Son of Shiva and Parvati

In Hinduism, Ganesha refers to the "lord of the hosts," also spelled as Ganesha and Ganesha, often also referred to as Ganapati)-and is one of the most well-known and venerated representations of God (Brahman).

He is the first born son of Shiva and Parvati, and the husband of Bharati, Riddhi and Siddhi. He is also called Vinayaka in Marathi, Malayalam and Kannada, Vinayagar and Pillayar (in Tamil), and Vinayakudu in Telugu.

'Ga' symbolizes Buddhi (intellect) and 'Na' symbolizes Vijnana (wisdom). Ganesha is thus considered the master of intellect and wisdom.

He is depicted as a big-bellied, yellow or red god with four arms and the head of a one-tusked elephant, riding on, or attended to by, a mouse. He is frequently represented sitting down, with one leg raised in the air and bent over the other. Typically, his name is prefixed with the Hindu title of respect, 'Shree' or Sri.

The cult of Ganesha is widely diffused, even outside of India. His devotees are called Ganapatya.The ganapati festival is celebrated with great enthusiasm and devotion throughout India but in Mumbai, the financial capital of the Country, the festival assumes a special significance because of the scale at which it is performed.

As is the case with every other external form with which Hinduism represents god, in the sense of the personal appearance of Brahman (also referred to as Ishvara, the Lord), the figure of Ganesha too is an archetype loaded with multiple meanings and symbolism which expresses a state of perfection as well as the means of obtaining it.

Ganesha, in fact, is the symbol of he who has discovered the Divinity within himself. Ganesha is the first sound, OM, in which all hymns were born. When Shakti (Energy) and Shiva (Matter) meet, both Sound (Ganesha) and Light (Skanda) were born.

He represents the perfect equilibrium between force and kindness and between power and beauty. He also symbolizes the discriminative capacities which provide the ability to perceive distinctions between truth and illusion, the real and the unreal. A description of all of the characteristics and attributes of Ganesha can be found in the Ganapati Upanishad (an Upanishad dedicated to Ganesha) of the rishi Atharva, in which Ganesha is identified with Brahman and Atman.

According to the strict rules of Hindu iconograpahy, Ganesha figures with only two hands are taboo. Hence, Ganesha figures are most commonly seen with four hands which signify their divinity. Some figures may be seen with six, some with eight, some with ten, some with twelve and some with fourteen hands, each hand carrying a symbol which differs from the symbols in other hands, there being about fifty-seven symbols in all, according to some scholars. The image of Ganesha is a composite one. Four animals, man, elephant, the serpent and the mouse have contributed to the makeup of his figure. All of them individually and collectively have deep symbolic significance.

The serpent is an archetype for DNA.

The Lord of Good Fortune

In general terms, Ganesha is a much beloved and frequently invoked divinity, since he is the Lord of Good Fortune who provides prosperity and fortune and also the Destroyer of Obstacles of a material or spiritual order. It is for this reason that his grace is invoked before the undertaking of any task (*e.g.* travelling, taking an examination, conducting a business affair, a job interview, performing a ceremony,) with such incantations as Aum Shri Ganeshaya Namah (hail the name of Ganesha), or similar. It is also for this reason that, traditionally, all sessions of bhajan (devotional chanting) begin with an invocation of Ganesha, Lord of the "good beginnings" of chants. Throughout India and the Hindu culture, Lord Ganesha is the first idol placed into any new home or abode.

Moreover, Ganesha is associated with the first chakra (energy wheel), which represents the instinct of conservation and survival, of procreation and material well-being.

Bodily Attributes

Every element of the body of Ganesha has its own value and its own significance:

- The elephant head indicates fidelity, intelligence and discriminative power;
- The fact that he has a single tusk (the other being broken off) indicates Ganesha's ability to overcome all forms of dualism;
- The wide ears denote wisdom, ability to listen to people who seek help and to reflect on spiritual truths. They signify the importance of listening in order to assimilate ideas. Ears are used to gain knowledge. The large ears indicate that when God is known, all knowledge is known;
- the curved trunk indicates the intellectual potentialities which manifest themselves in the faculty of discrimination between real and unreal; on the forehead, the Trishula (weapon of Shiva, similar to Trident) is depicted, symbolising time (past, present and future) and Ganesha's mastery over it;
- Ganesha's pot belly contains infinite universes. It signifies the bounty of nature and equanimity, the ability of Ganesha to swallow the sorrows of the Universe and protect the world;
- the position of his legs (one resting on the ground and one raised) indicate the importance of living and participating in the material world as well as in the spiritual world, the ability to live in the world without being of the world.
- The four arms of Ganesha represent the four inner attributes of the subtle body, that is: mind (Manas), intellect (Buddhi), ego (Ahamkara), and conditioned conscience (Chitta). Lord Ganesha represents the pure consciousness-the Atman-which enables these four attributes to function in us;
  - o The hand waving an axe, is a symbol of the retrenchment of all desires, bearers of pain and suffering. With this axe Ganesha can both strike and repel obstacles. The axe is also to prod man to the path of righteousness and truth;

- o The second hand holds a whip, symbol of the force that ties the devout person to the eternal beatitude of God. The whip conveys that worldly attachments and desires should be rid of;
- o The third hand, turned towards the devotee, is in a pose of blessing, refuge and protection (abhaya);
- o the fourth hand holds a lotus flower (padma), and it symbolizes the highest goal of human evolution, the sweetness of the realised inner self.

Ganesha is also defined as *Omkara* or *Aumkara,* that is "having the form of Om (or Aum) (see the section The names of Ganesha). In fact, the shape of his body is a copy of the outline of the Devanagari letter which indicates the celebrated Bija Mantra. For this reason, Ganesha is considered the bodily incarnation of the entire Cosmos, He who is at the base of all of the phenomenal world (Vishvadhara, Jagadoddhara). Moreover, in the Tamil language, the sacred syllable is indicated precisely by a character which recalls the shape of the elephant's head of Ganesha.

The broken tusk of Ganesha, as described above, stands primarily for his ability to overcome or "break through" the illusions of duality. However, there are many other meanings that have been associated with this symbol.

Ganesha and the Mouse

According to one interpretation, Ganesha's divine vehicle, the mouse or mooshikam represents wisdom, talent and intelligence. It symbolizes minute investigation of a cryptic subject. A mouse leads a clandestine life below the ground. Thus it is also a symbol of ignorance that is dominant in darkness and fears light and knowledge. As the vehicle of Lord Ganesha, a mouse teaches us to remain always on alert and illuminate our inner-self with the light of knowledge.

Both Ganesha and the Mooshak love modaka, a sweet dish which is traditionally offered to them both during worship ceremonies. The Mooshak is usually depicted as very small in relation to Ganesha, in contrast to the depictions of vehicles of other deities. However, it was once traditional in

Maharashtrian art to depict Mooshak as a very large mouse, and for Ganesha to be mounted on him like a horse. Yet another interpretation says that the mouse (Mushika or Akhu) represents the ego, the mind with all of its desires, and the pride of the individual.

Ganesha, riding atop the mouse, becomes the master (and not the slave) of these tendencies, indicating the power that the intellect and the discriminative faculties have over the mind. Moreover, the mouse (extremely voracious by nature) is often depicted next to a plate of sweets with his eyes turned toward Ganesha while he tightly holds on to a morsel of food between his paws, as if expecting an order from Ganesha. This represents the mind which has been completely subordinated to the superior faculty of the intellect, the mind under strict supervision, which fixes Ganesha and does not approach the food unless it has permission.

Married or Celibate?

It is interesting to note how, according to tradition, Ganesha was generated by his mother Parvati without the intervention of her husband Shiva. Shiva, in fact, being eternal (Sada Shiva), did not feel any need to have children. So Ganesha was born out of the exclusively female desire of Parvati to procreate. Consequently, the relationship of Ganesha and his mother is unique and special.

This devotion is the reason that the traditions of southern India represent him as celibate (see the anecdote Devotion to his mother). It is said that Ganesha, believing his mother to be the most beautiful and perfect woman in the universe, exclaimed: "Bring me a woman as beautiful as she and I will marry her."

In the north of India, on the other hand, Ganesha is often portrayed as married to the two daughters of Brahma: Buddhi (intellect) and Siddhi (spiritual power). Popularly in north India Ganesha is accompanied by Saraswati (goddess of culture and art) and Lakshmi (goddess of luck and prosperity), symbolizing that these qualities always accompany he who has discovered his own internal divinity. But this does not mean that Saraswati and Lakshmi are consorts of Ganesha. Symbolically this represents the fact that wealth, prosperity and success accompany those who have the qualities wisdom, prudence, patience, etc. that Ganesha symbolizes.

Mythological Anecdotes

How did He obtain His Elephant Head?

The highly articulated mythology of Hinduism presents many stories which explain the manner in which Ganesha obtained his elephant head; often the origin of this particular attribute is to be found in the same anecdotes which tell about his birth. And many of the these same stories reveal the origins of the enormous popularity of his cult.

Decapitated and Reanimated by Shiva

The most well-known story is probably the one taken from the Shiva Purana. Once, while his mother Parvati wanted to bathe, there were no attendants around to guard her and stop anyone from accidentally entering the house. Hence she created a boy's idol out of turmeric paste which she prepared to cleanse her body (turmeric was used for its anti-septic and cooling properties) and infused life into it, and thus Ganesha was born. Parvati ordered Ganesha not to allow anyone to enter the house and Ganesha obediently followed his mother's orders. After a while Shiva returned from outside and as he tried to enter the house, Ganesha stopped him. Shiva was infuriated at this strange little boy who dared to challenge him. He told Ganesha that he was Parvati's husband, and demanded that Ganesha let him go in. But Ganesha would not hear any person's word other than his dear mother's.

Shiva lost his patience and had a fierce battle with Ganesha. At last he severed Ganesha's head with his Trishula (trident). When Parvati came out and saw her son's lifeless body, she was very angry and sad. She demanded that Shiva restore Ganesha's life at once. Unfortunately, Shiva's Trishula was so powerful that it had hurled Ganesha's head very far off. All attempts to find the head were in vain. As a last resort, Shiva approached Brahma who suggested that he replace Ganesha's head with the first living being that came his way which lay with its head facing north.

Shiva then sent his celestial armies (Gana) to find and take the head of whatever creature they happened to find asleep with its head facing north. They found an elephant which slept in this manner and decapitated it, attaching the elephant's head to Ganesha's body and bringing him back to life. From then on, he was called Ganapathi, or head of the celestial armies and was to be worshipped by everyone before beginning any activity.

Shiva and Gajasura

Another story regarding the origins of Ganesha and his elephant head narrates that, once, there existed an Asura (demon) with all the characteristics of an elephant, called Gajasura, who was undergoing a penitence (or tapas). Shiva, satisfied by this austerity, decided to grant him, as a reward, whatever gift he desired. The demon wished that he could emanate fire continually from his own body so that no one could ever dare to approach him. The Lord granted him his request. Gajasura continued his penitence and Shiva, who appeared in front of him from time to time, asked him once again what he desired. The demon responded: "I desire that You inhabit my stomach."

Shiva granted even this request and he took up residence in the demon's stomach. In fact, Shiva is also known as Bhola Shankara because he is a deity easily propitiated; when he is satisfied with a devotee he grants him whatever he desires, and this, from time to time, generates particularly intricate situations. It was for this reason that Parvati, his wife, sought him everywhere without results.

As a last recourse, she went to her brother Vishnu, asking him to find her husband. He, who knows everything, reassured her: "Don't worry, dear sister, your husband is Bhola Shankara and promptly grants to his devotees whatever they ask of him, without regard for the consequences; for this reason, I think he has gotten himself into some trouble. I will find out what has happened."

Then Vishnu, the omniscient director of the cosmic game, staged a small comedy. He trasformed Nandi (the bull of Shiva) into a dancing bull and conducted him in front of Gajasura, assuming, at the same time, the appearance of a flutist. The enchanting performance of the bull sent the demon into ecstasies, and he asked the flutist to tell him what he desired. The musical Vishnu responded: "Can you give me that which I ask?" Gajasura replied: "Who do you take me for? I can immediately give you whatever you ask."

The flutist then said: "If that's so, liberate Shiva from your stomach." Gajasura understood then that this must have been no other than Vishnu himself, the only one who could have known that secret and he threw himself at his feet. Having liberated Shiva, he asked him for one last gift: "I have been blessed by you with many gifts; my last request is that everyone remember me adoring my head when I am dead."

Shiva then brought his own son there and substituted his head with that of Gajasura. From then on, in India, the tradition is that any action, in order to prosper, must begin with the adoration of Ganesha. This is the result of the gift of Shiva to Gajasura.

The Gaze of Shani

A less well-known story from the Brahma Vaivarta Purana narrates a different version of Ganesha's birth. On the insistence of Shiva, Parvati fasted for a year (punyaka vrata) to propitiate Vishnu so that he would grant her a son. Lord Krishna, after the completion of the sacrifice, announced that he would incarnate himself as her son in every kalpa (eon). Accordingly, Krishna was born to Parvati as a charming infant. This event was celebrated with great enthusiasm and all the gods were invited to take a look at the baby.

However Shani (Saturn), the son of Surya, hesitated to look at the baby since Shani's gaze is said to be harmful. However Parvati insisted that he look at the baby, which Shani did, and immediately the infant's head fell off and flew to Goloka. Seeing Shiva and Parvati grief stricken, Vishnu mounted on Garuda, his divine eagle, and rushed to the banks of the Pushpa-Bhadra river, from where he brought back the head of a young elephant. The head of the elephant was joined with the headless body of Parvati's son, thus reviving him. The infant was named Ganesha and all the Gods blessed Ganesha and wished Him power and prosperity.

Other Versions

Another tale of Ganesha's birth relates to an incident in which Shiva slew Aditya, the son of a sage. Shiva restored life to the dead boy, but this could not pacify the outraged sage Kashyapa, who was one of the seven great Rishis. Kashyap cursed Shiva and declared that Shiva's son would lose his head. When this happened, the head of Indra's elephant was used to replace it.

Still another tale states that on one occasion, the used bath-water of Parvati was thrown into the Ganges and this water was drunk by the elephant-headed Goddess Malini, who gave birth to a baby with four arms and five elephant heads. The river goddess Ganga claimed him as her son, but Shiva declared him to be Parvati's son, reduced his five heads to one and enthroned him as the *Controller of Obstacles* (Vigneshwara).

Ganesha the Scribe

In the first part of the epic poem Mahabharata, it is written that the sage Vyasa asked Ganesha to transcribe the poem as he dictated it to him. Ganesha agreed, but only on the condition that Vyasa recite the poem uninterruptedly, without pausing. The sage, in his turn, posed the condition that Ganesha would not only have to write, but would have to understand everything that he heard before writing it down.

In this way, Vyasa might recuperate a bit from his continuous talking by simply reciting a difficult verse which Ganesha could not understand. The dictation began, but in the rush of writing Ganesha's pen broke. He broke off a tusk and used it as a pen so that the transcription could proceed without interruption, permitting him to keep his word.

Sri Shiva Ratri—The night of Shiva:
vrndavanavani-pate! Jaya soma soma maule

sanaka-sanandana-sanatana-naradedya
gopisvara! Vraja-vilasi-yugangri-padme
prema prayaccha nirupadhi namo namaste

"O gatekeeper of Vrindavana! O Soma, all glories to you! O you whose forehead is decorated with the moon, and who is worshipable by the sages headed by Shanaka, Sanandana, Sanatana and Narada! O Gopisvara (Shiva), desiring that you bestow upon me that prema for the lotus feet of Sri Sri Radha-Madhava who perform joyous pastimes in Vraja-dhama, I offer obeisances unto you time and again."

Shiva (Lord Shiva) and ratri (night), so Shivaratri means the night of Lord Shiva. Every lunar month on the 13th or 14th day of the dark half of the moon phase there is a Shivaratri but in the Vedic month of Phalgun (February-March) there is a Maha Shivaratri. This year it falls on the 16th or 17th February depending where you are on the planet. The ceremony takes place mainly at night. This festival observed in the honour of Lord Shiva, who was married to Parvati (Mother Durga/ Mother Gauri) on this day.

Some dvevotees observe a strict fast on this day. Some really strict devotees perform a nirjal fast *i.e.* they do not even have a sip of water. They keep virgil the entire night. The Shiva lingam is worshipped throughout the night by washing it every three hours with milk, yogurt, ghee, honey and so forth, whilst chanting of the mantra "Om namah shivaya", continues. Offering of bael leaves are made to the Lingam. Bael leaves are very sacred (as Tulsi to Lord Vishnu and dhar to Mother Durga are very pleasing to Them) and Lord Shiva becomes pleased with that offering. Many different hymns glorifying Lord Shiva are sung with intense fervour and devotion. Aspirants perform japa of the Panchakshara mantra "Om namah Shivaya" (At the end of this article there is a very detail explanation of each syllable of this mantra.)

He who utters the names of Lord Shiva during Shivaratri with perfect devotion and concentration is freed from all sins that he/she have committed. He/she reaches the abode of Lord Shiva and resides very happily there.

Note: The Vaishnava/aspiring Vaishnava however, worships Lord Shiva on this day in order to receive His blessings to become a better devotee of Lord Krishna (Vishnu) and not to attain residence in the abode of Shiva. Vaishnavanams yatha Shambu-Lord Shiva is the greatest devotee of Lord Vishnu.

Lord Krishna (Vishnu) is in this analogy considered to be likened to Milk. Milk is the origin of so many dairy products- Kṛishna tu bhagavan swayam, and so Krishna or Vishnu is the origin of everything. When Krishna wants activity overseen of the modes of Tamo guna He expands Himself transforming His supreme powers to act in that way, as no-one else could, and in that state He is Shambu-Shiva. So it may be seen that milk that is transformed becomes yogurt, but that yogurt can never again beome milk, this is presented in Brahma samhita 5:45.

Brahma Samhita

ksiram yatha dadhi vikara-visesa-yogat
sanjayate na hi tatah prthag asti hetoh
yah sambhutam api tatha samupaiti karyad
govindam adi-purusam tam aham bhajami

Synonyms

ksiram—milk; yatha—as; dadhi—yogurt; vikara-visesa—of a special transformation; yogat—by the application; sanjayate—is transformed into; na—not; hi—indeed; tatah—from the milk; prithak—separated; asti—is; hetoh—which is the cause; yah—who; sambhutam—the nature of Lord Shiva; api—also; tatha—thus; samupaiti—accepts; karyat—for the matter of some particular business; govindam—Govinda; adi-purusam—the original person; tam—Him; aham—I; bhajami—worship.

Translation

Just as milk is transformed into curd by the action of acids, but yet the effect curd is neither same as, nor different from, its cause, *viz.*, milk, so I adore the primeval Lord Govinda of whom the state of Sambhu is a transformation for the performance of the work of destruction.

Purport

The real nature of Sambhu, the presiding deity of Mahesa-dhama, is described. Sambhu is not a second Godhead other than Krishna. Those, who entertain such discriminating sentiment, commit a great offense against the Supreme Lord. The supremacy of Sambhu is subservient to that of Govinda; hence they are not really different from each other.

The nondistinction is established by the fact that just as milk treated with acid turns into curd so Godhead becomes a subservient when He Himself attains a distinct personality by the addition of a particular element of adulteration. This personality has no independent initiative. The said adulterating principle is constituted of a combination of the stupefying quality of the deluding energy, the quality of nonplenitude of the marginal potency and a slight degree of the ecstatic-cum-cognitive principle of the plenary spiritual potency.

This specifically adulterated reflection of the principle of the subjective portion of the Divinity is Sada Shiva, in the form of the effulgent masculine-symbol-god Sambhu from whom Rudradeva is manifested. In the work of mundane creation as the material cause, in the work of preservation by the destruction of sundry ashuras and in the work of destruction to conduct the whole operation, Govinda manifests Himself as guna-avatara in the form of Sambhu who is the separated portion of Govinda imbued with the principle of His subjective plenary portion. The personality of the destructive principle in the form of time has been identified with that of Sambhu by scriptural evidences that have been adduced in the commentary.

The purport of the Bhagavata slokas, *viz.*, vaisnavanam yatha sambhuh, etc., is that Sambhu, in pursuance of the will of Govinda, works in union with his consort Durgadevi by his own time energy. He teaches pious duties (dharma) as stepping-stones to the attainment of spiritual service in the various tantra-shastras, etc., suitable for jivas in different grades of the conditional existence. In obedience to the will of Govinda, Sambhu maintains and fosters the religion of pure devotion by preaching the cult of illusionism (Mayavada) and the speculative agama-shastras. The fifty attributes of individual souls are manifest in a far vaster measure in Sambhu and five additional attributes not attainable by jivas are also partly found in him. So Sambhu cannot be called a jiva. He is the lord of jiva but yet partakes of the nature of a separated portion of Govinda.

On the chaturdasi of the waning phase of the moon in phalguna month occurs this festival of Shiva. By the mercy of Shiva, a devotee of the Lord, one can develop devotion to Sri Krishna. The vaishnavas do not follow the vrata of fasting and worship of Shiva at night, but simply request Shiva's mercy on this occasion. In Vrindavana many Gaudiya vaishnavas visit the temple of Gopesvara Mahadeva, where they offer flower garlands and pour ganges water on the Linga on this day.

Quotes from Vaishnava smriti compilation Hari Bhakti Vilas of Srila Gopal Bhattar Goswami, and the Dikdarshini Tika of Srila Sanatan Goswami Prabhupada.

Shiva ratri vrtam Krishna chatur-dasyantu phalgune
Vaisnaver api tat karyam sri Krishna pritaye sada
(Hari bhakti vilasa 14/187 from gautamiya tantra)

On the fourteenth day of the dark fortnight of the month of Phalguna (February-March), for the pleasure of Lord Sri Krishna, a Vaishnava should always take a vow to fast. (It is called Shiva Ratri, or the night of Lord Shiva.)

Parat parataram yanti narayana parayanah
Nate tatra gamisyanti ye dvisanti mahesvaram
(Hari bhakti vilasa 14/189 from kurma purana the supreme lord speaks to bhrgu Muni)

(One who avoids fasting on Shiva Ratri or the "night of Lord Shiva", he becomes very offensive. Whatever offences one accumulates by not fasting on Shiva Ratri day is explained here.) The destination of the Supreme Personality of Godhead, Lord Narayana, is certainly in the spiritual world. But if one is envious of Lord Sri Shiva, he does not attain the spiritual world.

Yani kany atra lingani sthavarani carani ca
Tesu sankramate devas tasyam ratrau yato harah
Shivaratris tatah prokta tena sa hari vallabhah

(Hari bhakti vilasa 14/200 from skanda purana, nagara khanda)

Whatever genital deities of Lord Shiva that can be found on this earthly planet, on the night of this fourteenth day of the dark fortnight of the month of Phalguna, Lord Shiva, who is a leading demigod, enters into them. For this reason, this day is called Shiva Ratri. For this reason, this day is very dear to Lord Sri Hari.

Sri Krishna vaisnavanantu prema bhakti vivardhate
Krishna bhakti rasa sara varsi rudra anukampaya
(Hari bhakti vilasa 14/221 from skanda purana)

Being a Vaishnava, if somebody fasts on Lord Shiva's night (Shiva Ratri day), by the mercy of him (Lord Shiva) who is diving in the ocean of the mellow of devotion to Krishna, one's devotion to Krishna increases rapidly.

The Story of King Chitrabhanu

In the Shanti Parva of the Mahabharata, Bhishma, whilst resting on a bed of arrows and discoursing on Dharma, refers to the observance of Maha Shivaratri by King Chitrabhanu. The story goes as follows.

Once, King Chitrabhanu was observing a fast with his wife, and it was the day of Maha Shivaratri. The sage Ashtavakra came on a visit to the king's court. The sage asked, "O king why are you observing a fast today?" King Chitrabhanu explained the reason, as he had the gift of remembering the incidents of his previous birth.

The king said to the sage, "in my past birth I was a hunter in Varanasi (Kashi). My name was Suswara. My livelihood was to kill and sell birds and animals. One day, I was overtaken by the darkness of night. Unable to return home, I climbed a tree for shelter. It happened to be a bael tree. I had shot a deer that day, but I did not have time to take it home. I bundled it up and tied it to a branch on the tree. As I was tormented by hunger and thirst, I kept awake throughout the night. I shed profuse tears when I thought of my poor wife and children who were starving and anxiously awaiting my return. To pass away time that night I engaged myself in plucking bael leaves and dropping them down onto the ground. "The day dawned and I returned home and sold the deer. I bought some food for myself and for my family. I was about to break my fast when a stranger came to me, begging for food. I served him first and then took my food."

"At the time of death, I saw two messengers of Lord Shiva, they were sent down to take me to his abode. I then learnt for the first time of the great merit I had earned but the unconscious worship of Lord Shiva during the night of Shivaratri. They told me that there was a lingam under that tree and that I had dropped the leaves on that Lingam. My tears, which I had shed out of pure sorrow for my family, fell onto the Lingam and washed it. And I had fasted all day and all night. Thus, I unconsciously worshipped Lord Shiva on that most auspicious day." "I lived in the abode of the Lord and enjoyed divine bliss for long ages. And now I am reborn as Chitrabhanu."

Lord Shiva's Assurance

When creation was completed Lord Shiva and Parvati Mata went to live on the top of Mount Kailash. Parvati asked, "O adorable lord, which of the many rituals observed in your honour pleases you the most." The Lord replied, "the fourteenth night of the new moon in the dark fortnight during the month of Phalgun, is my favourite day. It is called Maha Shivaratri. My devotees give me great joy and happiness by mere fasting than by ceremonial baths and offerings of flower, sweets and incense."

"The devotee observes strict spiritual discipline in the day and worships me in four different forms during each of the four successive three hour periods of the night. The offering of a few bael leaves is more precious to me than precious jewels and flowers. My devotee should bath me in milk in the first period, in yogurt in the second period, in ghee in the third, and in honey in the fourth and the last. In the morning he should feed Brahmins (that is if this is possible), and after doing that he/she breaks his/her fast. No ritual can compare with this simple routine in sanctity."

Parvati was deeply impressed by the words of her husband. She repeated this to her friends, who in turn passed it on to everyone else. Thus the sanctity of Maha Shivaratri was broadcast all over the world. The Shivaratri fast is aimed at controlling a persons two great vices (rajas–the quality of passionate activity) and (tamas-that of inertia).

The entire day is spent on worshipping Lord Shiva. Continuous worship of the Lord necessitates the devotee's constant presence in the place of worship. Ones motion is controlled and evils like lust, anger and jealousy, born of Rajas, are ignored and subdued. The devotee observes virgil throughout the night and thus conquers tamas also. Constant control is exercised over the mind. Every three hours a round of worship of the Shiva lingam is conducted and thus Shivaratri is the perfect vrat.

At many ashrams around the world Shivaratri is celebrated in the following manner.

Aspirants fast the whole day, many of them without taking a single drop of water. Some ashrams perform a grand *havan* (fire yajnas) to Lord Shiva for peace and welfare of all. The whole day is spent in chantinge of the mantra "Om namah Shivaya" and in meditation upon the Lord. At night all assemble in the temple and chant this mantra and during the four quarters of the night the Shiva lingam is worshipped with intense devotion.

Shiva Panchakshara Stotra

Stotra 1:

Nagendra Haaraaya Thrilochanaaya
Bhasmaanga Raagaaya Maheshvaraaya
Nityaaya Suddhaaya Digambaraaya
Tasmai Nakaaraaya Namah Shivaaya

Meaning: I offer my humble salutations to Lord Maheshvara-who has a garland of serpents around the neck; who has three eyes; whose body is covered with ash (vibhuti); who is eternal; who is pure; who has the entire sky as His dress and who is embodied as the first letter Na.

Stotra 2:

Mandaakini Salila Chandana Chaarthitaaya
Nandeesvara Pramatha Naatha Mahesvaraaya
Mandaara Pushpa Vahu Pushpa Supoojitaaya
Tasmai Makaaraaya Namah Shivaaya

Meaning: I bow to Lord Maheshvara, who is embodied as Makaara (letter Ma), whose body is anointed with holy waters from the river Ganges and sandal paste, who is the sovereign king of the Pramatha Ganas and who is adorned with innumerable divine flowers such as Mandaara.

Stotra 3:

Shivaaya Gauri Vadana Aravinda
Sooryaaya Dakshaadhvara Naashakaaya
Sree Neelakantaaya Vrisha Dhvajaaya
Tasmai Shikaaraaya Namah Shivaaya

Meaning: I offer my salutations to Lord Shiva, who is the resplendent sun for mother Gauri's lotus face, who is the destroyer of Daksha's sacrificial ritual, who is the blue necked Lord (due to the Haalahala poison which He agreed to consume), whose banner bears the emblem of a bull and who is embodied as the letter Shi.

Stotra 4:

Vasishta Kumbhodbhava Gautamaaya
Muneendra Devaarchita Sekharaaya
Chandraarka Vaishvaanara Lochanaaya
Tasmai Vakaaraaya Namah Shivaaya

Meaning: I prostrate before the God of Gods, who is worshipped and prayed to by great sages such as Vashishta, Agastya and Gautama, whose eyes are sun, moon and the fire and who is embodied as Vakaara (letter Va).

Stotra 5:

Yajna Swaroopaaya Jataadharaaya
Pinaaka Hasthaaya Sanaatanaaya
Divyaaya Devaaya Digambaraaya
Tasmai Yakaraaya Namah Shivaaya

Meaning: Prostrations to the sacred Lord who is the Yaksha incarnate, whose hairs are long and matted, who holds Pinaaka (trident) in His hand, who has the entire sky as His attire and who is embodied as the letter Ya.

Stotra 6:

Panchaaksharam Idam Punyam
Yah Pateh Shiva Sannidhau
Shivaloka Mavaapnothee
Shivena Saha Modate

Meaning: Whoever repeats this prayer composed with the five holy letters before Lord Shiva, attains that supreme abode of His and enjoys the eternal Bliss.

(nama shivaya)

The Linga Ashtakam (Eight verses glorifying the Shiva Linga.)

One chants these mantras when performing abhishekam. This ashtakam pleases Lord Shiva tremendously.

Brahma-murari surachit lingam, nirmal bhakchit shobhita lingam
Janma ja dukha vinashaka lingam, tat pranamami sadaa Shiva-lingam

I bow before that Sada-Shivalinga, which is adored by Brahma, Vishnu and revered by others, praised in mantras and which destroys the cycle of births and deaths.

Deva muni prawa-rarchita lingam, kamadahang karunakara lingam
Ravan darp vinashana lingam, tat pranamami sadaa Shiva-lingam

I bow before that Sada-Shivalinga, which is the focus of the worship of Saints, infinitely compassionate and the subduer of the pride and passion.

Sarva-sugandhi sulaypita lingam, budhi vivardhan kaarana lingam
Sidha-surasur vandita lingam, tat pranamami sadaa Shiva-lingam

I bow before that Sada-Shivalinga, smeared with Vibhuti, scented with perfumes and which elevates the power of the mind and kindles the light of discrimination, to which the Suras and Ashuras prostrate.

Kanak maha-mani bhukchita lingam, phani-pati veksh-tita shobhit lingam

Daksha suyagya vinashana lingam, tat pranamami sadaa Shiva-lingam

I bow before that Sada-Shivalinga, destroyer of King Daksh's ego, and which is decorated with various ornaments glowing within the embrace of the coiled serpent.

Kumkum chandana laypit lingam, pankaj haar sushobhita lingam

Sanchit paap vinashana lingam, tat pranamami sadaa Shiva-lingam

I bow before that Sada-Shivalinga, adorned with Saffron, Sandal-paste & garlands of lotus flowers, which exonerates all sinful actions.

Deva ganarchit sayvita lingam, bhavai bhakti bherewa-chaa lingam

Dinakar koti prabha kara lingam, tat pranamami sadaa Shiva-lingam

I bow that before Sada-Shivalinga, worshipped by the Sages & Saints with full faith and devotion and which is the source of unlimited brilliance.

Ashta dalopari veksh-tita lingam, sarva samudbhava kaarana lingam

Ashta daridra vinashana lingam, tat pranamami sadaa Shiva-lingam

I bow that before Sada-Shivalinga, destroyer of manifold wretchedness, which is the cause of creation and dissolution.

Soorguru soorwar poojit lingam, soorwan pushpa sadachita lingam

Paraatparam paramatmaka lingam, tat pranamami sadaa Shiva-lingam

I bow before that Sada-Shivalinga, the Transcendent Being, worshipped by the learned and all devotees, offering Bael leaves, flowers, fruits and water.

Lingaashtak medang punyang yuh, pathay-chiva sannidhow

Shiva-loka mawapanoti, shivena saha moditay

Whomsoever repeat these eight Slokas glorifying the Shivalinga in the divine presence of Lord Shiva attains final and everlasting relationship in the abode of Shiva Loka (if that's what you want and where you want to reside).

Demon Caikhacupoa Killed—taken from Krishna Book

Once upon a time, the cowherd men of Vrindavana, headed by Nanda Maharaja, desired to go to Ambikavana to observe the Shivaratri ceremony. The rasalila was performed during the autumn, and after that the next big ceremony is Holi, or the Dolayatra ceremony. Between the Dolayatra ceremony and the rasalila ceremony there is one important ceremony called Shivaratri, which is especially observed by the Shaivaites, or devotees of Lord Shiva.

Sometimes the Vainaavas also observe this ceremony because they accept Lord Shiva as the foremost Vainaava. But the function of Shivaratri is not observed very regularly by the bhaktas, or devotees of Kanaha. Under the circumstances, it is stated in Shrimad-Bhagavatam that Nanda Maharaja and the other cowherd men "once upon a time desired." This means that they were not regularly observing the Shivaratri function but that once upon a time they wanted to go to Ambikavana out of curiosity. Ambikavana is somewhere in the Gujarat province, and it is said to be situated on the river Saraswati. Yet we do not find any Saraswati River in the Gujarat province, although there is a river named Sabarmati. In India, all the big places of pilgrimage are situated on nice rivers like the Ganges, Yamuna, Saraswati, Narmada, Godavari and Kaveri. Ambikavana was situated on the bank of Saraswati, and Nanda Maharaja and all the other cowherd men went there.

They very devotedly began to worship the deity of Lord Shiva and Ambika. It is the general practice that wherever there is a temple of Lord Shiva, there must be another temple, of Ambika (or Durga), because Ambika is the wife of Lord Shiva and is the most exalted of chaste women. She doesn't live outside the association of her husband.

After reaching Ambikavana, the cowherd men of Vrindavana first bathed themselves in the river Saraswati. If one goes to any place of pilgrimage, his first duty is to take a bath and sometimes to shave his head. That is the first business. After taking a bath, they worshipped the deities and then distributed charity in the holy places. (SP-Krishna Book. Ch 34.)

Prabhupada: Observing Shivaratri, Nanda Maharaja and all the cowherds men, they increased their attachment for Kanaha. That is wanted.

(A.C.Bhaktivedanta Swami Prabhupada. 1st May 1974, Morning Walk excerpt, Bombay, India.)

"In this verse Lord Shiva is described as meohunoama, the best of the benedictors. He is also known as Acutona, which indicates that he is very quickly satisfied and very quickly angered. It is said in Bhagavad-gita that less intelligent persons go to the demigods for material benedictions. In this connection, people generally go to Lord Shiva, and because he is always quickly satisfied and gives benedictions to his devotees without consideration, he is called meohunoama, or the best of the benedictors. Materialistic persons are always anxious to get material profit, but they are not serious about spiritual profit.

Sometimes, of course, it so happens that Lord Shiva becomes the best benedictor in spiritual life. It is said that once a poor Brahma worshipped Lord Shiva for a benediction, and Lord Shiva advised the devotee to go to see Sanatana Gosvami. The devotee went to Sanatana Gosvami and informed him that Lord Shiva had advised him to seek out the best benediction from him (Sanatana). Sanatana had a touchstone with him, which he kept with the garbage.

On the request of the poor Brahma, Sanatana Gosvami gave him the touchstone, and the Brahma was very happy to have it. He now could get as much gold as he desired simply by touching the touchstone to iron. But after he left Sanatana, he thought, "If a touchstone is the best benediction, why has Sanatana Gosvami kept it with the garbage?" He therefore returned and asked Sanatana Gosvami, "Sir, if this is the best benediction, why did you keep it with the garbage?" Sanatana Gosvami then informed him, "Actually, this is not the best benediction.

But are you prepared to take the best benediction from me?" The Brahma said, "Yes, sir. Lord Shiva has sent me to you for the best benediction." Then Sanatana Gosvami asked him to throw the touchstone in the water nearby and then come back. The poor Brahma did so, and when he returned, Sanatana Gosvami initiated him with the Hare Kanaha mantra. Thus by the benediction of Lord Shiva the Brahma got the association of the best devotee of Lord Kanaha and was thus initiated in the maha-mantra, Hare Kanaha, Hare Kanaha, Kanaha Kanaha, Hare Hare/Hare Rama, Hare Rama, Rama Rama, Hare Hare.

All Good Qualities are in Shankara

Sri Sanatana Gosvami has written that some Vaisnavas don't observe Shiva Chaturdasi, the appearance day of Lord Shiva or Shankara. In Hari Bhakti Vilasa he has written that Vaishnava's should honour Shankara, and they may observe that date. I observe that date. I know Shankara, because in his form of Bankandi Mahadeva in Vrindavana he was the friend of Sanatana Gosvami. He became Bankandi Mahadeva from Gopisvara. Do you know this history?

When Sri Sanatana Gosvami became old, Shankara told him, "Now you are old. You should not come daily to see me, because you are coming from so very far away. Sanatana Gosvami replied, "I should come. I cannot change this habit." Then Gopisvara Mahadeva said, "Then I'm going to come very near to you–as Bankandi Mahadeva." He thus became Bankandi Mahadeva and resided very near the Madana Mohana Temple. Srila Sanatana Gosvami is also the very dear friend of Chakralesvara Mahadeva in Govardhana. He also used to be with Shankara in Kamyavan, where he is known as Kamesvara. He cannot be without Shankara.

As Nandisvara, Shankara has become the mountain of Nandagaon. He wanted that all Krishna's pastimes be performed on his back. Regarding Brahma, he has become Brahma-parvata in Varsana. Because he is so near to Radhika, he is therefore our Gurudeva These are the tattvas. We should try to honour Shankara as a great Vaishnava and as Guru. Don't dishonour him. We don't worship him separately, but we can observe Shiva Chaturdasi and glorify him in relation to his relationship with Sri Krishna. We should offer pranama to him with prayers like:

vrndavanavani-pate! Jaya soma soma maule
shanaka-Sanandana-Sanatana-naradedya
gopisvara! Vraja-vilasi-yugangri-padme
prema prayaccha nirupadhi namo namaste

"O gatekeeper of Vrindavana! O Soma, all glories to you! O you whose forehead is decorated with the moon, and who is worshipable by the sages headed by Shanaka, Sanandana, Sanatana and Narada! O Gopisvara, desiring that you bestow upon me that prema for the lotus feet of Sri Sri Radha-Madhav a who perform joyous pastimes in Vraja-dhama, I offer obeisances unto you time and again."

Shankara is telling Sri Narada that Prahlada Maharaja is superior to him. Why? Because he is "tricky". He does this only to give some encouragement to worldly persons. Actually, Prahlada cannot go to Vrindavana, but Shankara, as Gopisvara, resides there. As Hanuman, Shankara is always with Rama. As Bhima, he is with Krishna. When Hanuman and Bhima combine together in Kali-yuga, they become Madvacarya, our Sampradaya-guru. Thus, Lord Shankara serves in so many ways, and we should always honour him as our Guru. He is hundreds of thousands of times superior to Prahlada because he knows and meditates on asta-kaliya-lila. Parvati also meditates on asta-kaliya-lila. This is very secret, yet they both do it. Although Shankara is so much superior and more worshipful than Prahlada, still he says that Prahlada is superior.

Bibliography

Allen, Marc : *Tantra for the West: Everyday Miracles and other Steps for Transformation,* New World Library, San Rafael, 1992.

Anand, Margo : *The Art of Sexual Ecstasy: The Path of Sacred Sexuality for Western Lovers*, Jeremy P. Tarcher, Los Angeles, 1989.

Avalon, Arthur and Ellen : *Hymns to the Goddess*, Ganesh and Co., Madras, 1964.

Baba, Pagal : *Temple of the Phallic King : The Mind of India*, Simon and Schuster, New York, 1973.

Barrett David B. : *Seven Hundred Plans to Evangelize the World*, Birmingham, Alabama, U.S.A. 1988.

Basu, Manoranjan : *Tantras: A General Study*, Shrimati Mira Basu, Calcutta, 1976.

Bates, M. Searle : *Religious Liberty : An Enquiry*, New York, 1945.

Buchanan, Claudius : *Memoirs*, London, 1805.

Burke, Marie Louis : *Swami Vivekananda in America*, Calcutta, 1958.

Chuvin, Pierre : *A Chronicle of the Last Pagans*, Cambridge, Mass, USA, 1990.

Das, Sisir Kumar : *The Shadow of the Cross*, New Delhi, 1974.

David, M. D. : *Western Colonialism in Asia and Christianity*, Bombay, 1988.

Dayananda, Maharshi : *Satyarthaprakasha in Hindi*, Impression, Delhi, 1975

Desai, Mahadev : *Day to Day with Mahatma Gandhi*, Varanasi, 1968-1974.

Devi, Kamala : *The Eastern Way of Love: Tantric Sex and Erotic Mysticism*, Simon and Schuster, New York, 1977.

Dowman, Keith : *Masters of Enchantment*, Harper and Row, New York, 1988.

Downs, F. S. : *Christianity in North East India: Historical Perspectives*, Gauhati, 1983.

Duboi, Abbe : *Hindu Manners, Customs and Ceremonies*, Fifth Indian Impression, CUP, 1985.

Dyczkowski, Mark. S. G. : *The Canon of the Shaivagama and the Kubjika Tantras of the Western Kaula Tradition,* State University of New York Press, Albany, 1988.

Farrow, G. W. and Menon, I. : *The Concealed Essence of the Hevajra Tantra,* Motilal Banarsidass, Delhi, 1992.

Feuerstein, Georg : *The Philosophy of Classical Yoga,* Inner Traditions, Rochester, 1996.

Frawley, David : *Tantric Yoga and the Wisdom Goddesses,* Passage Press, Salt Lake City, 1994.

Gandhi, M. K. : *Christian Missions, Their Place in India*, Ahmedabad, 1941.

Gandhi, M. K. : *Hindu Dharma*, Ahmedabad, 1950.

Gandhi, M.K. : *The Collected Works*, 90 volumes, New Delhi, 1958 -1984.

Garrison, Omar : *Tantra: The Yoga of Sex*, Avon Books, New York, 1964.

Goel, Sita Ram : *Catholic Ashrams: Sannyasins or Swindlers?,* Voice of India, New Delhi, 1995.

Goel, Sita Ram : *Defence of Hindu Society*, Voice of India, Third enlarged edition, New Delhi, 1994.

Goel, Sita Ram : *Jesus Christ: An Artifice for Aggression*, Voice of India, New Delhi, 1995.

Goel, Sita Ram : *Papacy: Its Doctrine and History*, Voice of India, New Delhi, 1986.

Gopal, Sarvepalli : *Radhakrishnan: A Biography*, OUP, 1989.

Ishwar Sharan : *Jeevadhara : A Journal of Christian Interpretation,* New Delhi, 1988.

Ishwar Sharan : *The Myth of St. Thomas and the Mylapore Shiva Temple*, Voice of India, New Delhi, 1995.

Johnson, Paul : *A History of Christianity,* London, 1978.

Jyotirmayananda, Swami : *Vivekananda: His Gospel of Man-Making,* Madras,1988.

Keshavmurthi : *Sri Aurobindo: The Hope of Man*, Pondicherry, 1969.

Majumdar, R. C. : *The History and Culture of the Indian, People,* Bombay, 1981.

Mangalwadi, Vishal : *Missionary Conspiracy: Letters to a postmodern Hindu*, Mussoorie, U.P., 1996.

Mitter, Partha : *The Much Maligned Monsters*, Oxford, 1977.

Nehru, Jawaharlal : *Glimpses of World History*, OUP reprint, Fourth Impression, 1987.

Neill, Stephen : *History of Christianity in India: The Beginnings to 1707 AD,* Cambridge University Press, 1984.

Panikkar, K. M. : *Asia and Western Dominance*, London, 1953.

Panikkar, K. M. : *Malabar and the Portuguese*, Bombay, 1929.

Panikkar, Raimundo : *The Unknown Christ of Hinduism*, London 1964.

Plattner, Felix Alferd : *The Catholic Church in India; Yesterday and Today*, Allahabad, 1964.

Pope John Paul II : *Crossing the Threshold of Hope*, New York, i994.

Price, Rev. J. Frederick : *The Private Diary of Anand Ranga Pillai*, Madras; 1904.

Priolkar, A.K. : *The Goa Inquisition (Bombay, 1962),* Voice of India, New Delhi, 1991.

Ram Swarup : *Introduction to Anirvan?s Inner Yoga (Antaryoga)*, New Delhi, 1988.

Ram Swarup : *Introduction to D. S. Margoliouth"s Mohammed and the Rise of Islam*, New Delhi, 1985.

Ram Swarup : *Pope John Paul II on Eastern Religions and Yoga: A Hindu-Buddhist Rejoinder*, Voice of India, New Delhi, 1995.

Ram Swarup : *Ramkrishna Mission in Search of a New Identity*, Voice of India, 1986.

Ram Swarup : *The Word As Revelation: Names of Gods*, New Delhi, 1980.

Shiva Rao, B. : *The Forming of India?s Constitution*, Bombay, 1967.

Shiva Rao, B. : *Transfer of Power 1942-47: The Quit India*, London, 1971.

Singhal, D. P. : *India and World Civilization*, Calcutta, 1972.

Temple, Richard : *Oriental Experience*, London, 1883.

Temple, Richard : *Thirty-four Conferences Between the Danish Missionaries and Malabarians (or Heathen Priests)*, London, 1719.

Temple, Richard : *United Nations and Human Rights, The, UNO*, New York, 1984.

Thomas, M. M. : *The Acknowledged Christ of the Indian Renaissance*, Madras, 1976.

Thomas, P. : *Christians and Christianity in India and Pakistan*, London, 1954.

Vivekananda, Swami : *The Complete Works*, Calcutta, 1985 impression.

Vivekananda, Swami : *Vatican Council II: The Conciliar and Post Conciliar Documents*, Bombay, 1988.

Young, Richard Fox : *Resistant Hinduism*, Vienna, 1981.

Index

F

G

H

I

J

K

L

M

N

O

P

R

S

T

U

V

W

Y

❑❑❑